Volume 8

THE CHANGING STRUCTURE OF
THE WORLD OIL INDUSTRY

THE CHANGING STRUCTURE OF THE WORLD OIL INDUSTRY

Edited by
DAVID HAWDON

Routledge
Taylor & Francis Group

LONDON AND NEW YORK

First published in 1985 by Croom Helm Ltd

This edition first published in 2018
by Routledge
2 Park Square, Milton Park, Abingdon, Oxon OX14 4RN

and by Routledge
711 Third Avenue, New York, NY 10017

Routledge is an imprint of the Taylor & Francis Group, an informa business

British Library Cataloguing in Publication Data
A catalogue record for this book is available from the British Library

ISBN: 978-1-138-10476-1 (Set)
ISBN: 978-1-315-14526-6 (Set) (ebk)
ISBN: 978-1-138-30972-2 (Volume 8) (hbk)
ISBN: 978-1-138-30979-1 (Volume 8) (pbk)
ISBN: 978-1-315-14348-4 (Volume 8) (ebk)

Publisher's Note
The publisher has gone to great lengths to ensure the quality of this reprint but points out that some imperfections in the original copies may be apparent.

Disclaimer
The publisher has made every effort to trace copyright holders and would welcome correspondence from those they have been unable to trace.

The Changing Structure of the World Oil Industry

**Edited by
David Hawdon**

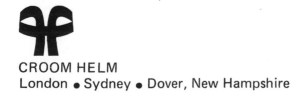

CROOM HELM
London • Sydney • Dover, New Hampshire

© 1985 David Hawdon
Croom Helm Ltd, Provident House,
Burrell Row, Beckenham, Kent BR3 1AT

Croom Helm Australia Pty Ltd, First Floor,
139 King Street, Sydney, NSW 2001, Australia

British Library Cataloguing in Publication Data

The Changing structure of the world oil industry.
   1. Petroleum industry and trade
   I. Hawdon, David
   338.2'7282   HD9560.5

   ISBN 0-7099-3717-2

Croom Helm, 51 Washington Street,
Dover, New Hampshire 03820, USA

Library of Congress Cataloging in Publication Data
Main entry under title:

The Changing structure of the world oil industry.

   "Based on contributions to the second international
energy economics conference held at the University of
Surrey in March 1984" — Acknowledgements.
   1. Petroleum industry and trade — Congresses.
I. Hawdon, David. II. Title.
HD9560.5.C46 1985 338.2'7282 84-23769
ISBN 0-7099-3717-2

Printed and bound in Great Britain by
Biddles Ltd, Guildford and King's Lynn

CONTENTS

ACKNOWLEDGEMENTS

The papers included in this volume are based on
contributions to the second international energy
economics conference held at the University of
Surrey in March 1984. Judi Pollard and Jan Benn of
the University's Bureau of Industrial Liaison were
responsible for the pleasant atmosphere and
efficient administration of the conference. The
financial assistance of the U.K. Department of
Energy in enabling academic and government
economists to participate is gratefully
acknowledged. Preparation of the book was
facilitated by the excellent transcription services
of the Arab-British Chamber of Commerce. Sally
Silverman provided the expert word processing and
secretarial input needed to produce the final text.

INTRODUCTION AND SUMMARY

David Hawdon

Throughout its history the economic benefits
conferred by the international petroleum industry
have sometimes been seen as flawed by certain non-
competitive characteristics of its organisation.
Monopolistic features of the industry have, at
various times, been held responsible either for oil
prices being set higher than costs, however defined,
or for restraint in oil production designed to
ensure stability in price levels. Thus John D
Rockefeller in the 1890s achieved control of the US
oil industry by backward integration from refining
to crude production by using discriminatory transport practices combined with undoubted technical
superiority over his rivals. At the present time
there is a trend in the policies of certain Arab oil
producers towards vertical integration by the
acquisition of downstream operations such as
transport and refining facilities. This suggests
that the concentration of economic power in the
industry, by integration or other means, not only
has been perceived by producers as important but
will continue to be a necessary if not always
sufficient condition for attaining their objectives.

A brief analysis of the major features of
petroleum industry market structure is in order as
background to the detailed issues examined in the
specialist papers which were originally presented at
an international energy economics conference in the
University of Surrey in March 1984 and have now been
completely revised. Perhaps the most widely
discussed phenomenon is the degree of vertical
integration exhibited within the industry. The
dominance until 1973 of the industry by eight very
large integrated operators combining crude oil
production, refinery and marketing functions within
each firm is well known, as are the underlying

economic and physical factors which encourage such consolidation. The high ratio of capital to variable costs in refining provides an incentive to maximise refinery runs and stimulate a demand for secure dependable supplies.The long lines involved in moving oil from wellhead to refinery together with the high costs of storage, act as spurs to the substitution of administered planning for competitive markets at each stage (except of course transport itself which has remained competitive and in chronic excess supply save for several short run crises). Finally, the natural forces of oil production resulting in substantial periods of easily expanded supply and low incremental costs lead at times to attempts to control output by the extension of control over the various stages of the industry.

It is important to note that vertical integration of itself is not a sufficient condition for market distortion. The crucial factors are the effects of vertical integration on the costs of the industry, the existing industry structure and the objective functions of the firms. Suppose total costs were to fall through integration, this might have no effect on output if marginal costs were to remain unchanged and the firm was a profit maximising monopolist. On the other hand, should the firm's objective be that of constrained revenue maximisation, the integration would tend to produce an expansion of output. Where integration results in no change in costs, however, and there is a fixed proportional link between output at various stages in the industry, there will be no effect on output. This is because any previous degree of monopolisation at any stage will already have been taken into account in the cost functions of firms at succeeding stages and in their output decisions, and hence in the industry output.

More important from the point of view of impact on oil industry output and price levels have been the horizontal links between vertically integrated producers. These took the form of joint ventures in e.g. Kuwait (Kuwait Oil Company) and Saudi Arabia (Aramco) and in the pre 1973 period of long term contracts between the leading producers. In the case of joint ventures, with the exception of the K.O.C., output restraint was achieved by the requirement for each participant to disclose its "ordering out" plans together with the fact that most of the majors were present in more than one venture and hence were able to acquire significant

information on intentions of rivals without the need
for explicit collusion. Further restrictions were
achieved by the tendency for crude short companies
to tie up surplus oil supplies from rivals in long
term contracts. This hindered any expansion in oil
supplies by the crude short companies and promoted
the interests of both parties.

The market power of the major companies has of
course been replaced by that of a non integrated
producer grouping - namely OPEC - largely since
1973, although this was the culmination of a process
begun by the Tehran-Tripoli agreements of 1971. On
the other hand the basic factors favouring
integration remain and it is little wonder that
various attempts have been made by the oil producing
countries to integrate forward in attempts to secure
markets for final products. These attempts are
understandable in the light of increased pressure on
the price of crude oil, expanded output by non OPEC
producers and developments in downstream markets.
The various papers which follow attempt to assess
these developments and to determine their likely
consequences for the oil market.

In a personal account of the history of the
oil industry Lord Kearton, formerly head of the
British National Oil Corporation, gives his
reactions to the major developments. His paper
illustrates the early links between oil and
chemicals and the failure of potential competition
from coal to oil conversion processes in the 1920s.
Stresses between host governments and the majors are
traced back into the 1930s even though they posed
little threat to the majors' dominance of the market
until much later. The attempt by the producing
countries to replace the companies as the
controllers of oil in the 1970s is seen as only
partially successful, and the remarkable resilience
and adaptability of the companies is emphasized.
Finally the role of government policy towards the
development of publicly owned oil companies
operating in the North Sea area is described and
assessed and stress is laid on the transfer of
technology and information to the new state
corporations. OPEC's influence is seen as resting
on the strength of Saudi Arabia and on the interests
of the consumer countries in higher levels of oil
prices to protect their energy investments. The
reduction of the companies' power is seen as
beneficial in general for the future development of
the industry.

Paul Stevens, in a well documented survey

paper, argues that integration within the international oil industry enabled producers to keep oil
prices consistently above marginal costs throughout
the post 1945 period despite the existence of
significant excess production capacity. Statistical
measures of excess capacity based on the concept of
easily brought forward production are developed
which make allowance for nominal inventories and
excludes capacity development which might have
adverse effects on economic development. The
maintenance of such excess capacity is explained by
reference to political factors as well as by the
desire of existing producers to restrain price
competition. They were able to restrain competition
through the joint venture system until the sixties,
but even after 1973 when joint ventures were
replaced by host country owernship of crude, the
majors continued to act as lifters of crude and
hence could prevent production exceeding their
downstream and arms length demand. OPEC's action in
establishing producion quotas in 1982 is seen as an
attempt to restore control when the former system
finally collapsed. However the market has now lost
the informational benefits of company integration
which make the task of policing its members'
behaviour more difficult for the OPEC cartel.

The consequences of the loss of equity
producing interest in the OPEC countries for the
largest of the major oil companies are investigated
by Richard Reid, President of Esso Europe. The loss
of crude capacity together with changed prospects of
sales growth and uncertainty about oil prices have
led companies to intensify their search for new
reserves and to adopt a profit centre approach to
company operation. For Exxon, downstream activities
face greater political and economic risks and have
now to be justified in terms of current market
information rather than longer term considerations.
Governments have a role to play in stimulating the
primary upstream activities of exploration and
development by appropriate oil taxation policies.

In the next paper Jack Hartshorn analyses the
role of government oil traders in a market which has
become increasingly dominated by arms length or
third party sales. The Iranian revolution and the
second oil price shock which followed, destroyed
almost all vestiges of the old system of
preferential prices for the integrated majors.
Instead government sellers now rely upon the spot
market for information on differentials and make
their profits on crude sales only. Their motives

differ from those of the companies which they
replaced – in particular their rates of social time
discount are lower and they have little incentive to
develop resources outside their own frontiers.
These factors have encouraged a greater expansion of
refinery construction than would have occurred in a
private market. The OPEC producers have substantial
incentives to seek stability in prices and share
this aspiration with the integrated majors. The
role of the futures market although often involving
only paper barrels at the moment is likely to assume
greater significance over time.

Ian Seymour examines the impact of the
structural changes on OPEC's attitudes and
objectives. 1982 is seen as a significant watershed
when OPEC abandoned its traditional objective of
improving oil prices in favour of one of demand
recovery through reduced prices. A new weapon –
production pro rating – is now available to help
prevent prices falling uncontrollably but may prove
to be of less significance than the price discipline
which arises out of the need to preserve revenues.
The costs and benefits to OPEC countries of
diversifying downstream into refining and retail
distribution are assessed and emphasis is laid on
the shortage of investment funds and management
skills as limiting factors in this process. In the
marketing sphere, sales of packages of various
crudes have become ways of coping with the
restrictions on differentials in the 1983 London
agreement. This option is not, however, available
to all OPEC producers and the problem of
differentials is likely to grow. The benefits of
the new realistic OPEC approach to policy making are
emphasized.

The problems of predicting developments
downstream when the overall industry structure has
changed so radically and is indeed still in a state
of disequilibrium are analysed by Edith Penrose.
Many of these changes resulted from the break up of
the integrated oil companies, but the actual changes
– the replacement of long term by short term
contracts, the dominance of spot market trading, and
the rise of a futures market – were as much the
product of specific political developments as of
alterations in industry structure. The downstream
market has been affected on the one hand by the
increase in the number of crude suppliers and on the
other by the heightened uncertainty of oil supplies.

Future trends in oil industry structure and
stability, together with their implications for

Introduction and Summary

North Sea oil and gas development are discussed by a
panel consisting of Peter Beck, David Howell MP,
Walid Khadduri, Colin Robinson and John Wiggins.
During the conference the opportunity was taken to
survey opinions as to the future course of crude oil
(marker) prices and the results of this survey are
reported in the final section in order to provide a
record of price expectations at a time of continuing
uncertainty regarding the future development of the
oil market.

1 THE OIL INDUSTRY. SOME PERSONAL RECOLLECTIONS AND OPINIONS

Lord Kearton

INTRODUCTION

I must say that it does seem odd that someone who
retired a few years ago should be making the opening
remarks at this conference on the changing structure
of the world oil industry. Especially as that
someone, me, was never employed in any of the major
or even minor companies of the industry, and is not
an economist, not a business historian or an analyst
or a planner. I did, it is true, intermittently
come into contact with the world of oil over a
period of about 50 years and what I propose to do
is to give you a sort of worm's eye reaction. And I
have lately, been a member of an energy sub-
committee of a select committee of the House of
Lords; so I am not a complete innocent, although
far-removed from the distinguished experts who will
be offering the meat of this conference. And for
that matter, the distinguished participants in the
conference. But the meat provided by the experts
will be spiced with the insight of a minister, I am
very pleased to see, an ex-Secretary of State.

Well, I had better start off by saying I
regard the oil industry with admiration. And I have
also had occasion to view it with some suspicion.
Admiration for its boldness, its risk-taking and the
ever-growing technical ingenuity and engineering
expertise which it exercises. Suspicion because of
the industry's very bigness and capacity to have
undue influence on governments and their decisions
and even to over-awe governments. The industry has
had its share of ups and downs. In its up periods
it has often acted very independently and self-
sufficiently with little regard to any interests but
its own. When in difficulties, it has been very
adept at enlisting government support for
restrictive practices. Up or down, it has always

1

favoured and mostly operated in managed markets.
Competition has been within in a carapace of
cartelisation.

Obviously the structure of the industry has
evolved and changed as oil became the single most
important commodity in the world economy. Nevertheless it is my feeling that the instincts and
behaviour which have characterised the industry for
more than a century still remain, although modified
by the experiences of the past decade. Will it be
that beyond the convolutions of today, we shall find
it a case of "plus, ca change, plus c'est la même
chose"?

INDUSTRY BEGINNINGS
Rockefeller in the early days of the industry set
the pattern. He of course was a book-keeper. We
call them accountants nowadays. He was the founding
father of the big oil company ethos. That is quite
straight forwardly, control the supply of the raw
material, control the markets for the finished
products, control prices, and cartelise. His
Standard Oil Trust through its operations was the
main begetter of the anti-trust laws in the United
States. Yet his big oil was tiny in comparison with
today's operations. The main outlets for oil in his
day were for lighting and for cooking. The takeover
by oil of the energy and transportation markets was
fifty years into the future. But oil as a comfort
and convenience commodity was so important at the
turn of the century that government intervention to
establish a freer market was inevitable. Governments have sparred with oil companies ever since.
It has not been a love hate relationship so much as
a need hate relationship, like a stormy marriage
with peaceful intervals.

The government-inspired fragmentation of the
old Standard Oil was, ironically, helpful to the
industry itself. Opportunities for growth were so
great that they outran the capacity of Standard's
monolithic organisation. The rump of the diverse-
ment directives, Standard Oil of New Jersey, now
Exxon, today takes turn and turn about with General
Motors to be the largest business enterprise - in
terms of turnover - in the Western World. The
offshoots themselves became giants. And Texas grew
two of its own - Gulf and Texaco.

America was where the oil industry began and
the big oil company was born. Finds in Russia and
the Dutch East Indies led to the founding of Shell
and Royal Dutch at the turn of the century, at first

2

in competition with Standard in Europe and with each other, and led by entrepreneurs as ruthless as Rockefeller. Shell and Royal Dutch amalgamated in the early years of the century. And also before the First World War, the discovery of oil in Persia led to the growth of a company which ultimately became BP, another world-class enterprise. For nearly fifty years, a handful of large companies developed and controlled oil fields all over the globe. The companies of this handful were rivals, but also collaborators. Together, they had more economic power and influence than all but the largest of the nation states. They shaped and changed our lives.

WORKING FOR ICI IN OIL
Turning from such wide vistas to personal experience, I had my own introduction to the oil scene in the 1930s. I joined ICI to work on their project to make oil, particularly petrol, from coal. In the 1920s, oil production was but 5 per cent or less of today's production. Nevertheless, some shortages had developed and interest had been aroused in processes to convert coal into oil. One route, developed by the German chemist, Bergius, proceeded by hydrogenation and high pressure. Shell and Standard had worked on similar processes. Bergius was backed by the vast German chemical concern, I.G. Faben-Industrie. The formation, in 1926, of ICI from an amalgamation of four long-established British chemical companies was partly, even mainly, inspired by the wish to participate in the new process. In other words, the thought in the mid-20s was that making petrol from coal was going to be the great growth area for the big chemical companies. International Hydrogenation Patents was formed with Shell, Esso, I.G. and ICI as partners, pooling their respective expertise.

Neither of the oil majors in the event went ahead with actual hydrogenation plants. More oil was being found, shortages were turning into glut. And now much more important to the oil companies, Shell and Esso had with BP arrived at the secret Achnacarry Agreement in 1928. This divided the world into an international cartel. The three oil companies eschewed competition and agreed on market shares. It was envisaged that new capacity would be brought on to the market in an orderly way. They had merely to maintain a watching brief on oil from coal, and the activities of the chemical companies they once feared.

ICI built its coal-oil plant in 1933/34 in a

3

mood of hope and desperation. With desperation, because the slump of that time had left them with lots of spare hydrogen making capacity which they wanted to use. Hope, that an oil shortage would again develop. Germany went ahead because it wanted oil self-sufficiency in the event of war.

Fifty years ago I worked on various aspects of the ICI project and became the Manager of the refinery section. It was also one of my jobs to watch the general supply-demand situation. Now, at the time, two things struck me. In America, production in the 1930s was strictly controlled to protect prices. The rationing was carried out by an authority of that most free-wheeling and individualistic of States, Texas. The State brought order into the market and protected the producers. Moreover, import of cheap oil into the USA was strictly controlled, to protect the home producers.

In the United Kingdom I, from my refinery, supplied petrol to identical specification to both Shell and Esso. It was a time when rival advertising between Shell and Esso was very fierce, very extensive. The consumer was given the impression that these two famous names were fighting for his custom, that the petrols offered were different. They were not different. They were exactly the same. And the market was being shared out anyway.

In the late 1930s with the threat of war looming, the British government became concerned with the security of aviation fuel supply for the RAF. I was one of the ICI team seconded to come up with proposals. It was decided to provide a petrol base by hydrogenating gas oil and to bring this up to the grade required by the aviation engines by blending with Iso-octane and adding lots of tetra-ethyl lead. ICI pioneered the production of the Iso-octane from Iso-butane, a by-product of the hydrogenation process. The process was licenced from the consultants Universal Oil Products of Chicago, who had operated a small pilot plant. I was put in charge of ICI's Iso-octane operations, through design to start-up. I also served as a technical manager in the tripartite team from Shell, Trinidad Leaseholds, and ICI which designed the large-scale aviation fuel plant which was built at Heysham. It was the largest single war-time project, incidentally, in which ICI was engaged. Trinidad supplied the gas oil and some of the general management. The major technical input by far was from ICI and this was true even for the

4

distillation columns of the plant where one might
have expected Shell to score.

In 1939, to widen my background, I was given
the opportunity to visit plants of Esso, Shell, Gulf
and others in the United States covering both
refining and reforming and chemicals production. In
passing, I also saw numerous oil fields, all of the
nodding donkey type. At this time, Esso was by far
the most impressive oil company in the technical
sense. I was at Baton Rouge, and saw the early days
of fluidised bed development and of catalytic
cracking.

The most disconcerting thing to me about the
tour was the monumental indifference of the American
oil men to the outcome of any possible war in
Europe. You must remember that this was 1939 and we
had had the crisis at Munich and war did seem a
possibility. The American oil men were extremely
indifferent to any outcome and, if anything, Germany
was favoured.

The self-absorption of oil men was illustrated
to me again in 1940 when Holland was overrun and a
number of Shell technical men from the Hague escaped
to this country and were temporarily housed by ICI.
It was one of my jobs to look after them. The
minutest points of procedure about office amenities
and fittings seemed of far more importance than
anything else that was going on in Europe.

At the end of 1940 I got moved on to atomic
work and stayed on atomic work for nearly all the
war except for a brief period when I was brought
back to work on a project called "Victane" which,
believe it or not, produced aviation fuel with a
specification of 125 octane number and enabled the
Tempest air plane to overtake and overturn the
flying bombs.

POSTWAR CONTACTS WITH THE INDUSTRY, TO 1960

Post war I joined Courtaulds and again took part in
sundry dialogues with various of the major oil
companies. The activities included operation of a
pilot plant in Houston with Esso, prior to a joint
project which in the event did not materialise. It
also extended to discussion of various collaborative
proposals with Shell, again none of which were
pursued. The impression left on me was that these
very powerful companies, Shell and Esso, could not
adjust their thinking to collaboration on a basis of
equality with a weaker partner. I think that in the
1950s the OPEC countries, where the stakes were so
vastly greater, must have felt the same way.

5

The Oil Industry

In the oil production world resentment had of course started earlier. Mexico reacted angrily against what it regarded as exploitation and nationalised its oil industry in 1938. It was boycotted and it took over 30 years to make the State Company, Pemex, efficient. I had the opportunity of seeing Pemex in the late 70s and it seemed to me to be a good operation. I see their results for 1983 show a five billion dollar profit. To come back to my theme, Iran nationalised BP in 1951 and then found its oil boycotted by agreement amongst the major oil companies. It was 1954 before a new agreement was reached which allowed bulk exports again from Iran. This agreement, reached with American help, ended BP's monopoly of Iranian oil. Marketing was put in the hands of a new consortium made up besides BP of American and French interests with BP retaining a 40 per cent share. Iran was left owning the oil, but output, distribution and pricing were firmly in the hands of the oil companies.

It was in 1960 that Exxon provided the trigger which led to the formation of OPEC. It unilaterally cut the price of Venezuelan crude for the second time in quick succession. And this had repercussions, not only on Venezuela, but on Iraq and the whole Middle East and caused great anger.

Now Exxon's price reduction was actually sparked off by cuts in the price of Russian oil, especially to Signor Mattei in Italy. I knew something of Mattei because in the 50s I represented my then company Courtaulds in various Italian interests, and one of them was in fact a buyer of natural gas from Mattei. He was a folk hero in Italy in those days. His Agip petrol stations, with their logo of the six-legged running dog with flames coming out of its mouth, really made the whole Po Valley look very colourful. Mattei was very anti big oil. I believe he was the man who coined the phrase "the Seven Sisters".

I had a personal reason to be interested. From the mid-50s, I had worked hard to open up UK trade with Russia. I sold them fibres, fibre manufacturing plants, and fabrics and with some success. I might mention that trade has continued and we recently celebrated in my old company 30 years of trade with Russia. And the total trade over the 30 years has amounted in today's money to about 1.6 billion pounds. So we started something quite useful in the 50s. But the Russians at the time were short of hard currency. I looked for

6

further ways of increasing trade, and discussed with
the Russian authorities the idea of importing their
oil and oil products. They were very keen on the
idea. And with the Russians we looked at and worked
out various schemes. The initial phase was to be
the importation of one million tonnes per year of
oil. I was very pleased. We took the proposals to
Whitehall for blessing. I got a blank refusal. Not
just blank, but passionate blank. I was told
categorically that no circumstances could be
envisaged in which approval could be given. My
arguments that the oil would be cheaper and broaden
the supply base were peremptorily dismissed. The
then permanent Secretary concerned told me that any
such imports would be over his dead body. You must
forgive me for seeing the hands of the major oil
companies in his attitude.

Well, perhaps it was just as well for my
company Courtaulds that I was turned down. It would
have been a formidable task to have taken on the oil
majors, even if modestly. And the venture would
have been far away from our main line of business.
It was conceived as an exercise in diversification,
then a fashionable concept.

OPEC, formed in 1960, was not at first
effective. The 60s saw the oil majors dominating
the world oil scene. Together they controlled over
70 per cent of the world trade. Remember, the 60s
was a marvellous period for world economic growth.
The fields in the Middle East and North Africa
seemed easy to find, the wells were extraordinarily
prolific, capital costs were modest and actual costs
of production almost derisorily low. But cheap as
was the delivered price of oil to the West, it could
have been even cheaper if a free market had
existed. But production was controlled by the
various licencees and they had agreements with each
other which crossed all national boundaries. If a
producing country was difficult it found it harder
to get its oil uplifted. Unity within OPEC was
poor. The Arab boycott at the time of the 1967
Egypt-Israel six-day war was not effective.
Venezuela and Iran quickly made up any shortfall.

Please note the scenario at this time. For
years and years, oil consumption had continued to
grow between 8 and 10 per cent compound growth per
annum. Forecasts by the oil companies at the end of
the 60s assumed the curve would continue upwards at
this and faster rates throughout the 70s the 80s and
so on. The companies were confident the oil would
be there. The plateau forecast for Saudi Arabia,

its potential, was put at 20 million barrels a day.
And it was thought that other large reserves would
be found. The oil concessionaires saw as their
chief task not the finding of more oil, but the
maintenance of an orderly market for the oil already
available. They wanted an oil price which gave
"reasonable" - their definition - margins over
production and all other costs, including payments
to host countries. These host countries were
becoming increasingly restive. Very much so under
this regime. They felt their cut from the margin
was too small. They wanted both a higher price and
a bigger cut from the increased margin.

Looked at in one way, the non-American
countries with indigenous oil were being exploited
by the multi-national countries because of their
oil. Looked at in another, the exploration,
production, marketing and stable pricing policies of
these same multi-nationals were vital factors for
over 20 years in the extraordinary economic growth
of the whole world system.

A feature of oil company structure from
earliest days has been vertical integration, right
from the days of Rockefeller. Controlling outlets,
as I mentioned, was very important. The big
companies had their own refineries and distribution
systems. They controlled a lot of their own
transport. Refineries and their conversion
techniques were developed in the home of petroleum
technology, America. Initially, some big multi-
nationals preferred to keep their refineries away
from both producing and importing countries. I
remember, long ago, the then Shell chairman arguing
in one of his annual reports that the logical place
for large multi-product refineries was sites such as
the Dutch West Indies. Near enough to take the
Venezuelan crude, and big enough to supply a wide
variety of export markets with a wide variety of
products - including countries such as Britain. But
the rapid growth in consumption led of course in the
end to the establishment of versatile refineries in
most countries which consumed large amounts of oil.

The end of the 60s, this wonderful period, saw
major developments. The United States, a protected
market for so long, became a net importer of oil and
oil products. They could no longer meet their own
demands and this was a great turning point. The
tidy world of the Seven Sisters was also being
disrupted by brash independents, themselves growing
into major international oil companies. Then
Colonel Gaddafi staged his coup in Libya and this

8

was absolutely one of the seminal events of the
whole oil industry development. He showed in
essence the Emperor had got no clothes. He showed
in essence that the West, when it came to it, would
not protect what had been thought to be their vital
interests. The restlessness of the OPEC countries
began to strain at the reins of the multi-natioal
control system. As the 70s opened, the emancipation
struggle strengthened. It culminated at the time of
the 1973 Yom Kippur war. This saw the OPEC
countries finally breaking free, and taking decisive
control of the pricing of their oil. The wider
consequence was that the vulnerability of the
prosperity of the developed world was vividly
demonstrated.

The oil multi-nationals, in the event, came to
terms with the new situation surprisingly rapidly.
They still controlled the markets. It became clear
that for many years the OPEC countries would need
them just as before. In 1977, I took part in a big
meeting at Vienna between the OPEC countries and
representatives of all the major oil companies.
They had not really met in the open for four years,
since 1973. Very wide-ranging discussions and a lot
of feeling out for positions took place at this
meeting and by the end of it, it was quite clear a
certain rapprochement had taken place. You must
remember that although the oil had been
nationalised, there were no financial disasters in
the oil companies. Their balance sheets had never
capitalised the value of reserves in the ground. On
revenue account, cash flow shot up with the price of
oil, and profits on crude, on a much smaller "cut",
were still very attractive. "Old oil" in non-OPEC
states became much more valuable. New oil, in
Alaska and the North Sea, and elsewhere round the
world, could now tolerate greatly increased capital
expenditure per barrel and still yield handsome
profits, handsome dividends.

BNOC

The exploitation of Alaska and the North Sea
depended on overcoming enormous technical problems.
To begin with, the know-how was overwhelmingly
American. But it rested far more in the staffs of
the specialist companies serving the oil industry
than in the oil companies themselves. I went to the
ceremonial opening of the Forties field in 1975 and
the Queen and the rest of the Royal Family were
there. I must say BP put on a magnificent show. It
was a British oil field and a British company.

9

Actually, ninety-five per cent of the engineering
and development was done by Americans. But it was
not deemed very sensible to say so at the time. It
was in 1975 that I became associated with North Sea
activities in helping to set up BNOC - The British
National Oil Corporation. In the second half of
that decade I witnessed an enormous increase in the
in-house capabilities of the major oil companies.
The specialist companies grew too, but the end
result was far more of a partnership in technology
development than formerly. Britain greatly
benefitted from a build-up of design and
manufacturing skills within its own resources - a
build-up greatly encouraged by direct government
guidance and pressure. The Offshore Supplies Office
played a major part. In the latter half of the 70s
BNOC also played a part but an enormous amount
really did depend on the persistence and push of the
then Secretary of State, Mr Wedgwood Benn. I am
very pleased to see that today's Department of
Energy, and Ministers such as Mr Buchanan Smith, are
following his example.

The Southern gas fields of the North Sea were
discovered and exploited in the mid and late 60s.
The supply contracts with the state monopoly buyer
were not linked to the price of oil. At the end of
the 60s, the oil companies were convinced that the
price of oil would continue to fall - which is
extraordinary looking back! - and they made 25-year
contracts both here and in America with low fixed
prices for gas. What they wanted was a stable floor
price for the gas. In this country this meant that
when oil prices shot up the windfall profits on gas
passed directly to the State, via the British Gas
Corporation. The early oil licences were very
favourable to the oil companies. All the bargaining
strength was on their side. The United Kingdom
government departments they dealt with at that time
had very few specialist staff and at first lacked
the will as well as the resources to negotiate too
strongly. Before 1973, it was felt that the UK
needed the oil company skills, commitment and money
much more than the oil companies needed North Sea
oil.

There was no sudden conversion on either side
after 1973. Although the multinationals had had to
concede oil sovereignty to the OPEC countries they
saw no need to concede it to the United Kingdom
government. On the UK government's part, the value
to the economy of getting the North Sea developed
precluded any strong-arm or expropriation tactics.

The Oil Industry

I remember suggesting to Mr Healey that he could be
bolder in dealing with the oil companies. He said
he couldn't take the risk of the oil companies
leaving the North Sea. I said that if the North Sea
was ringed with cavalry and tanks, it wouldn't keep
the oil companies out. And his reply was: even if
there were only a 1 per cent risk they were going to
pull out, he couldn't take it and he didn't. So,
persuasion was used to alter the situation.
        First, the taxation anomalies were corrected,
for which we owe a great debt to Mr Edmund Dell.
Then a 51 per cent equity participation was
negotiated for all existing and new oilfields but on
a basis which cost the country nothing and left
control of the operations in the hands of the
companies themselves and the equity profits before
tax also remained in company hands. What the
participation deals did was to give secure control
of destination and of prices. At the same time, the
operating arm of the new national oil company of
which I was the first chairman and chief executive
developed its own capabilities in the field of
exploration, drilling, platform design and
operation. At one time BNOC was participating in
over 40% of all drilling in the North Sea. This was
in a period when the oil companies were sulking for
some reason. BNOC also showed its capability for
independent fund-raising. We got into terrible
trouble with a Select-Committee of Parliament
because we went off to America and raised I think it
was $850 million with no specific government backing
and no security except our promise that we would
produce the oil and pay the tenders back in due
time. Later BNOC showed its capability in selling
large quantities of oil internationally. The
important thing really about all this was that the
myth that the necessary skills to develop the North
Sea and the financial resources to back it were
unique to the existing majors was gently dismantled.
The information flow to the Department of Energy and
other government departments was greatly increased.
We started to advise the Foreign Office on oil
matters including the situation of oil around the
Falklands - this was long before the Falkland crisis
came. I recall our advice was that there wasn't
more than a 1 in 50 chance of there being commercial
oil there - still, that's by the way. But before
BNOC came along, the Foreign Office depended very
much on what Shell and BP chose to tell them. It is
important to say that during this period of the late
70s the Department of Energy and other Whitehall

11

departments very considerably strengthened their staff, both specialist and general, to deal with oil matters. The balance, from being biased, was made even between the parties.

In the early days the concept, and even the very existence of BNOC was bitterly attacked by oil interests, both large and small. And just for good measure they attacked people associated with it such as myself. Later, the new Corporation was accepted. I remember going to a conference in Geneva in 1979 and Mr Pocock, the then Chairman of Shell was a member on a panel, and he said to an international audience, "I think I am going to surprise you all by saying I have come to the conclusion that BNOC is rather a good thing". So, by 1979, BNOC had become accepted by the oil companies. I would be a little worried if this acceptance proceded to assimilation.

BNOC was welcomed of course by other state oil companies. Cordial relationships were developed with the Norwegians, the Brazilians, the Venezuelans, the Mexicans, the Malaysians, the Indians, Nigerians and countries of the Middle East. BNOC got quite close to Kuwait for instance. And contacts began with Saudi Arabia. Exchanges of technical staff were initiated with Venezuela and Malaya. Advice was given to the Chinese in dealing with the multi-nationals – and advice was also given to the Danes. And in the field of oil sales close relationships were established with a worldwide range of customers.

In the turnaround in policy which followed the 1979 election in Britain, the changes – I say this with great trepidation in the presence of the Minister of State – have been cosmetic to some extent. The participation element of BNOC has been retained in toto. With the current oil glut, the value of the participation deals in establishing secure access to oil is temporarily overshadowed. The present usefulness lies in helping to smooth out oil price fluctuations and to allow a unified approach to world oil price discussions. The partial privatisation of the operating arm of BNOC, now Britoil, has not meant a complete hands off policy. The Government has directors on the board, it has a so-called golden share, it keeps a very close eye on Britoil. And the existence of Britoil is a potent reminder to the oil companies that they are not, indeed, indispensable. If circumstances change, a full governmental role for Britoil could be easily reactivated.

The Oil Industry

The British approach to its indigenous oil, as
it has developed, is both relaxed but firm. I think
it is generally accepted that it is fair. The one-
sidedness of ten years ago, in favour of the oil
companies, has gone. It seems to me that the set up
is now sufficiently flexible to deal with any
foreseeable eventualities, without a crisis in
industry-government relationships. It remains, like
so many things British, sui generis.

OIL IN THE 80s

It is interesting that the situation facing the oil
industry in the 80s is the same as in the 60s. How
to cope with the supply capability which for some
time ahead will exceed demand. The policy which has
been adopted is exactly the same as that developed
by the State of Texas in the 1930s - pro-rationing.
A central body decides which oil fields shall
produce what and reviews the market, and the
problems of individual producers, every few months.
If necessary, prices are adjusted, including
relative prices, and market shares are modified.
The individual producers lobby the controlling body,
as in the 1930s.

The controlling body today of course is the
OPEC Council of Ministers. It has not the statutory
force which the old Texas Railroad Commission
enjoyed. It is voluntary. It is a producers'
cartel, but one in which the pressures of individual
producers are very different. It is a cartel with
many participants - far more than when a handful of
multi-nationals decided matters. It is a cartel
which holds prices far far above the cost of
production of any of the producers. It is a cartel
whose successful functioning is vital to the very
economic existence of most of the members, which is
why it has worked, and why I think it will continue
to work. Even so, it has been the will power, the
example and the capacity for ruthlessness of Saudi
Arabia which has been decisive in its success.
Could one hazard that the producing countries
learned well from big oil?

The OPEC countries have the satisfaction, or
the chagrin, of holding their price umbrella over
all other producers. All - the western world,
communist world, developing worlds alike. And these
other sources of supply can happily let the oil flow
to the physical limit of the wells, governed only by
good oil field practice. To no country does this
protection matter more than to Great Britain. A
sudden and sharp fall in the price of oil would blow

13

away the projections of the Chancellor's recent
Green Paper and multiply his difficulties.
Conversely, it would aid Britain's main international competitors. The Chancellor must be very
relieved that he won his poker game with OPEC a year
ago, when he was Secretary of State for Energy.

Should lower oil prices be a major objective
for the world? Would it be a good thing for the
world if OPEC collapsed? A year or so ago many
economists were giving very positive answers to
these questions - Yes. I do not notice the same
certainty as of today.

It is true the world recession of the past few
years has indeed been largely blamed on the oil
price explosions of the 70s. There is no doubt that
the higher price of energy was a considerable
factor. However, the sustained economic boom of the
previous 30 years was due to a great many favourable
conjunctions, uniquely so. I have not got time to
list them all. And some of these had run their
course and were petering out during the 70s. In the
present world circumstances their re-establishment
seems unlikely. A lower price of oil would not in
itself rekindle the growth rates of the past. On
the other hand, a price of oil sharply above present
price levels would strangle at birth the incipient
recovery in world trade.

What is the position of big oil companies on
oil prices? I have the feeling they are happy with
the present situation. They work closely and
constructively with the OPEC countries. As I have
said, both groups need the other. The influence of
the oil companies is directed towards ensuring that
any changes in price are gradual. There is no overt
sign that any individual multi-national wants to
rock the boat, and gain perhaps a temporary
advantage by destablising the market. There is
natural and justifiable self-interest in this.
Prospecting in such areas as the Beaufort Sea or the
China seas needs an assurance of floor prices.

Assuming there is price stability at roughly
present levels for some time ahead, what of the
future? Against all the forecasts once made so
confidently, oil consumption has not grown in the
past five years, but has actually declined.
Notwithstanding, oil still remains the most
important single commodity in the world. Assurance
of long-term supplies is vital. And there are some
signs that the decline in consumption may be over
and that demand may be beginning to grow again. It
is not in the world's economic interests that former

growth rates should be resumed. There is less
confidence than there was fifteen years ago that
resources would always expand to keep pace.

Back in the 1930s, when ICI watched somewhat
despairingly for an oil shortage to develop, the
ratio of reserves to annual production seemed to
stay at about 30. Fifty years later, at levels of
production some 25 times greater, that ratio is much
the same. While vast new areas have not been
located in the past decade and a half, smaller
accumulations, such as the North Sea, have turned
up.

I remember attending a World Power Conference
in 1977 at which papers were presented claiming that
oil still to be discovered and won was about twice
what the world had consumed so far. And great play
was made with the reserves locked up in tar sands,
heavy oil belts, shale deposits and so on. Interest
in the latter sources has waned since '77. Oil
which flows or can be pumped is so much more
attractive. The oil company budgets for traditional
exploration have increased and the most hostile
environments are under examination. I have the
feeling that if the 1977 forecasts of a lot of extra
oil to be discovered really come close to
realisation, then despite the wide ranging search,
most of the new reserves might still be in the
Middle East.

Interest in coal liquefaction - which is where
I came in - must always involve hydrogen addition in
some form or another. It had a revival in the
'70s. This interest has died away. Demonstration
projects have been scaled down or abandoned. The
House of Lords sub-committee I mentioned earlier
thought it important that a research activity be
maintained, helped by European Community funds, but
recognised liquefaction was very much for the longer
long-term.

GENERAL COMMENTS

You will hear from the experts their views as to the
future structure of the oil industry. I can only
offer a worm's eye comment or comments. Firstly,
the giants of the last fifty years remain the giants
of the industrialised world both relatively and
absolutely. When one sees that Exxon's 1983
turnover was about 95 billion dollars with after-tax
profits of about 9 billion dollars; when one sees
that Shell's turnover was about 75 billion dollars,
their after tax profits were about 7 or 8 billion
dollars and so on, they really are huge enter-

prises. Their revenues outstrip those of many
nations, not just the tiddlers of the post-colonial
era, but quite important nations. The big oil
companies continue to grow organically and by
acquisition. Some of the acquisitions are
substantial. The bid by Mobil for Superior Oil, the
biggest of the so-called independents in America,
amounts to 5.7 billion dollars. The bid by Texaco
for the international Getty Oil is almost double
this at 10.1 billion dollars. And most striking of
all is the recently announced bid by one of the
Seven Sisters for another - Socal's proposed merger
with Gulf. BP transformed its own position a few
years ago by its tie-up with Sohio which it now
controls. So I think these huge conglomerations of
influence and power and potential will remain.
Secondly, while the industry swarms with tiny, with
small and with medium-sized oil companies, I do not
think any will develop into an eighth sister. But
they do have a most important function. They bring
liveliness and some unorthodoxy into the industry.
The technical and pioneering boldness of Philips in
the Ekofisk development in the North Sea is a case
in point. The willingness to wildcat on hunch has
unlocked much treasure both in the past and in
recent times. One notices that if these second rank
companies develop beyond a certain point they tend
to be swallowed up, if not by a sister, then by
other giants, in the way Conoco went to Dupont.
Thirdly, the industry has always attracted and
produced men of remarkable and diverse ability. A
recent example is Boone Pickens. A geologist who
started out less than 30 years ago with about two
and a half thousand dollars has become rich as fast
as Paul Getty did. His Mesa Petroleum has been
skilful at finding, developing and selling gas. It
also found oil in the Moray Firth, an area which
most other companies thought had very poor
prospects. I met Boone Pickens because BNOC bought
his Moray Firth interests, his Beatrice field, from
him. He sold us Beatrice at a price of a dollar a
barrel in the ground. Fortunately he sold it to us
before the 1979 price explosion, so for us it turned
out a good buy. Now, it is interesting that
Pickens' forays in the shares of other oil companies
have been even more dynamic. He has demonstrated in
recent months that even a Sister can be stalked.
The Gulf-Socal marriage is a reaction to his
mobilisation of shareholder influence, and the way
in which a company can be attacked for managing its
affairs perhaps not in the best interest of share-

holders. Now fourthly, the major companies - and I think this is most important - are now again concentrating on their main justification - finding and developing oil accumulations. There is still an interest on their part in coal. But exercusions into nuclear power, into metal and mineral deposits, into general retailing and office equipment are no longer popular. Fifthly, national oil companies are here to stay. I think all countries with indigenous oil now have a national oil company, except the United States and Great Britain. And even Great Britain has half of one. Sixthly, the major multi-nationals have adjusted to the national oil companies, accept them as a part of life, and work with them. The initial anguish and fears of the companies vanished as they found they could even work through them. The OPEC countries still depend enormously on their partnerships and arrangements with the various multi-nationals so long attracted.

The companies themselves are far more conscious than they were of national susceptibilities. They deal more carefully with governments and with public opinion. They have been successful in projecting a better image. They no longer give the impression of over-aweing. But they do remain supra-national entities with power and influence largely independent of so-called democratic control. While they no longer deal with governments as masters, they still do so as equals, and it is in an equality now clothed in respectability.

I conclude by saying that in my view the independent oil companies, vast, large, medium and beginners are essential, absolutely essential to the growth and prosperity of the industrialised world. Their past contributions have been enormous and will remain so in the future. But I welcome the growth of a parallel capability on the part of national interests. Power, whether presidential, prime ministerial or private, should always be offset by a system of checks and balances. I feel sure the big multi-nationals now accept this philosophy and, who knows, they may even welcome it.

2 A SURVEY OF STRUCTURAL CHANGE IN THE INTERNATIONAL OIL INDUSTRY 1945-1984

Paul Stevens

1. INTRODUCTION

The purpose of this paper is to present an historical survey of changes in the industrial structure of the international oil industry (IOI) since 1945. The IOI refers to the world outside of the Communist Bloc and North America (WOCANA). The subject of industry structure, according to the economic textbooks, covers a variety of aspects such as the size of firms, the ease of industry entry and the elasticity of demand for output, etc. Industry structure is usually considered in terms of its relationship to the conduct of firms with respect to their objectives, price setting behaviour and their attitudes to rivals. The intention in this paper is to concentrate on only two aspects of industry structure, namely integration in two forms, horizontal and vertical. This is not to say that other aspects of structure or conglomerate integration are unimportant or uninteresting but requirements of time and space must limit the coverage.

Much has been researched, written and discussed on integration in the IOI, as can be seen from the reference lists at the end of the paper, although horizontal integration has received much less attention than has vertical integration probably as a result of the divestiture debate which has taken place in the USA. Because much of the material is well known, rather than simply go over familiar ground the paper will link the discussion of structure to a specific hypothesis although the hypothesis is by no means original. Since 1945, and many would argue since 1908 and the D'Arcy discovery in the Middle East, the IOI has been characterized by two elements. The first is very large potential excess crude oil producing capacity coupled with

very low marginal cost. Given these two conditions
in an oligopolistic structure, conventional economic
wisdom would predict a downward pressure on prices
and probably intense price rivalry. Since 1945,
such pressure has been evident as is the case at
present. However, it has occurred much less than
might have been expected and the history of the
period has been interspersed with two massive step
jumps in price. The hypothesis of this paper is
that the lack of price competition for crude oil for
most of the period can be explained in terms of the
integrated structure of the industry and the way in
which that integrated structure has changed. It is
towards substantiating this hypothesis that the
survey of industry integration is directed. The
issue is more than just of interest to the historian
of the industry. Much of the present situation of
the IOI and its future can be explained and analysed
in terms of how the integrated structure of the
industry has changed. The Marquis of Halifax once
remarked that the best qualification for a prophet
was a good memory. It is with that in mind that
this paper has been written.

The paper is divided into five sections.
After this introduction the second section examines
why the industry structure is important. This is
intended to provide empirical justification for the
earlier assertions of excess capacity and low
marginal cost. The third section examines the way
in which the horizontally integrated structure of
the industry controlled the excess supply up until
the early seventies and then what subsequently
replaced the structure. The fourth section examines
the role of vertical integration in the context of
excess capacity and low marginal cost and the final
section is a conclusion.

2. WHY INDUSTRY STRUCTURE IS IMPORTANT
In the context of the paper, the importance of
industry structure arises from two characteristics
of the IOI, excess capacity and low marginal cost.
This section of the paper is to expand upon these
points and to examine their consequences.

2.1 Excess Potential Capacity
The concept of crude producing capacity is
relatively straightforward. Technically there are
various stages from wellhead to terminal coupled
with the flow rate from the pools. Maximum capacity
is given by the weakest link in that chain. However, translating the concept into an empirical

19

reality is extremely complex and fraught with practical and political problems (Martyniuk, 1983; Wildavsky & Tenebaum, 1981). Firstly, there is the distinction between the installed capacity and the maximum sustainable capacity which in turn is a function of the maintenance standard of the equipment and the rate sensitivity of the fields. For example, an estimate for OPEC capacity in 1981 suggests an installed capacity of 40.5 million b/d but a maximum sustainable capacity of 35.2 million b/d (Martyniuk, 1983). Secondly, there are the constraints upon the use of capacity. Leaving aside for the moment the 'willingness' of the crude owner to produce, two major constraints can be identified. The first, relevant only since the early seventies, is the presence of facilities to utilize the associated gas coupled with the willingness to flare. Secondly, since no one wants crude for its own sake, consideration must be given to the downstream capacity to transport, process and distribute. This can get extremely complex. It has been argued that if 'cheap' crude is available then there is an incentive for the development of downstream capacity. However, it is unlikely that an investor would risk large quantities of capital on downstream expansion unless the likely returns were substantial. Certainly in refining returns seem generally to have been low or even negative although as will be discussed later this is the result of transfer pricing within the vertical integration of the industry. Thus 'only fools and affiliates pay posted prices'.

In this paper, when crude capacity is being considered it is really potential crude capacity which is under review rather than existing installed capacity. This distinction gives rise to the type of statement common in the literature pre 1970. For example, 'outside the USA it is doubtful whether there is a real surplus of physical capacity ... but at current prices enormous amounts of crude in proved reserves could readily be developed to afford a comfortable return' (Hartshorn, 1966, page 366. See also Adelman, 1972, page 162).

Assuming for the moment it is possible to obtain some estimate of potential crude producing capacity what then constitutes excess capacity? At first sight it would appear to be potential capacity less production. This however is misleading. Part of this 'excess' would be for seasonal variation although the use of annual data should overcome this problem. Part will also be for contingencies. Above

20

ground crude storage has always been an expensive
exercise and it has probably been cheaper to invest
in excess producing capacity than in storage tanks.
One source suggests that in the sixties in the
Middle East and Venezuela some 15 percent of
capacity was required as insurance (Adelman, 1972,
page 163). Taking out these elements one is then
left with genuine surplus capacity but even this is
not the end of the story. Part of that surplus may
not be 'available' since there may be physical
blocks on its production such as war or there may be
unwillingness on the crude owners to produce. The
reasons for 'unwillingness' can be complex when
decision makers are governments. On the one hand
neo-classical economic theory asserts that so long
as marginal revenue exceeds marginal costs then the
producer is irrational not to produce. Leaving
aside the issue of future versus present prices in
the case of a depletable resource well covered in
depletion theory (Hotelling, 1931), there are social
costs and benefits associated with oil production.
These might lead to output decisions different from
those which one would make given price expectations
and discount rates. For example, if the marginal
utility of revenue is zero or even negative because
of its 'damage' to the society then this can lead to
backward bending supply curves (Cremer & Isfahani,
1981), a diagrammatic expression of 'unwillingness'. Nor does such neo-classical analysis say
anything about political decisions not to become
vulnerable to pressure by accumulating assets in
other countries as has been the case for a number of
oil producers. It is only when all these elements
are removed that one is left with a surplus
potential capacity which is held back specifically
in order to restrain competition in the disposal of
crude oil which is the surplus capacity of interest
to this paper.

While the whole problem of converting the
concept into an operationally useable tool is
fraught with difficulty, the following attempt has
been undertaken. There is no time series data for
installed producing capacity in WOCANA let alone
potential producing capacity and even cross
sectional data on installed capacity is very limited
before the late seventies. Therefore a rough
methodology has been developed in an effort to provide some estimate of potential capacity. Potential
capacity is defined as production which with a
little effort could be brought forward and produced
now rather than at some time in the future from

21

existing proven reserves. This is not unreasonable since it is generally accepted that additional production from many WOCANA sources could have been brought onstream very rapidly (Adelman, 1972, page 217). One has only to note the speed with which the loss of Iranian production in 1951 was made up from the other countries to provide some evidence for this assertion. Indeed the history of the industry before the seventies was littered with fields which having been discovered were closed in with no development, possibly one of the most controversial being North Rumaila.

Two options have been chosen reflecting different lead times. In both, 1968 is taken as a watershed. In option 1 before 1968, potential capacity by region and in some cases countries Yt is defined as the production level in Yt + 2. After 1968, the potential capacity is defined as production one year hence (i.e. Yt + 1). The reason for the choice of 1968 as the 'watershed' is two-fold. First, in many areas after this date (approximately) oil tended to be found and developed in more inaccessible areas and therefore more effort would be needed to produce earlier. This was in part due to the growing uncertainty over the availability of more accessible Middle East crude. This uncertainty also provides a second reason for the use of 1968 as a 'watershed'. The date marks the start of growing uncertainty over control of crude oil on the part of the companies following the emergence of the 'participation' issue (Stevens, 1976) and hence there was a moratorium on the development of any new capacity in many regions.

So far the methodology presents no problems provided that production is forever growing but what if in some years production falls? Here two criteria apply. If the production reserve ratio remains above twenty years then the previous peak capacity remains as the capacity. If however this ratio falls below twenty years then the capacity falls either to match the production of that year or five percent of the proven reserves whichever is lower. This explains the occasional figure which suggests production capacity utilization above 100 percent.

Option 2 mirrors option 1 except the lag is extended to three years before 1968 and two years after. The magnitude of the lag is somewhat arbitrary. For example there is little doubt that vast producing capacity could be onstream in the Middle East with limited effort thus a lag of 5-10

years would not be ridiculous. If pressed for an
opinion the author feels that the most realistic
option is a combination of the two, 3 years before
1968 and 1 year afterwards although that still
understates Middle East potential capacity. Once
potential producing capacity has been determined
(Yt), production (Pt) is subtracted to obtain excess
capacity (Et).

The results of the analysis are presented in
tables 1 and 2. This shows the total WOCANA excess
capacity as a percentage of total capacity by the
two methods together with the regional origins of
the excess capacity. While the aggregate excess
capacity is different between the two options as can
be seen from figure 1, the regional source of the
excess varies little between the methods as can be
seen from figures 2 and 3. Figures 2 and 3 also
show that the excess capacity is dominated by the
big four middle east producers and Libya. Of the
big four's excess capacity between 1958-1970, 38
percent was in Iran, 29 percent in Saudi Arabia, 20
percent in Kuwait and 14 percent in Iraq. Of the
African excess capacity on average over the whole
period 68-74 percent was in Libya, 9-10 percent was
in Nigeria and 15-22 percent was in the rest of
Africa depending on the option chosen. Finally, it
is worth noting the exponential rise in excess
capacity after 1979. So far, the empirical
'attempt' to quantify excess capacity supports what
is generally known and accepted.

Certain factors suggest that the results are
understated and other factors suggest they are
overstated. There are five elements which
contribute to the possible exaggeration of the
excess capacity. First, the figures do not take
account of physical non availability because of war
or 'conservation' decisions. Figure 4 is an attempt
to point out some of the major losses from this
source. Secondly, nothing is said about
maintenance. For example, Kuwait's capacity
according to the methodology has been at 3.03
million b/d since 1971. However, it is unlikely
this figure could now be reached without considerable investment. Similarly, Iran has an alloted
capacity of 6.04 million b/d since 1973. Following
the 1979 revolution and the Gulf War this figure is
certainly a gross overstatement. Thirdly, the
excess capacity includes contingency 'storage' which
is not strictly speaking surplus capacity. Fourth,
it takes little or no account of the decline rate of
fields apart from the rather rough rule of thumb on

production reserve ratios. To account for this the potential capacity should have been computed on a well by well basis since there 'may be wide variation among wells in a different resevoir' (Adelman, 1972, page 60). However, the methodology in the paper is too rough to make such sophistication worthwhile and data would not be available. However, the purpose is to measure potential rather than actual capacity and so the production reserve ratio rule is not completely without foundation.

Finally, there is an overstatement which derives from the methodology. In a region where production is rising very rapidly then this will show considerable excess capacity simply because of the lag effect. If considerable investment and effort is going into production expansion then it may be questionnable to assert that the production levels in two or one year's time could have been produced now. In that sense the potential capacity is not strictly speaking surplus. One has the feeling that much of the apparent African excess capacity in the sixties was the result of this methodological quirk.

To offset the overstatement, two factors suggest that the figures understate the excess. Firstly, the figures are derived from annual data which disguises wide variations in production even allowing for seasonality. For example the Saudi Arabian 1979 capacity is given as 9.9 million b/d derived from the 1980 production figure of 9.9 million b/d. Yet in the fourth quarter of 1980 production reached an average of 10.5 million b/d (Petroleum Economist). While there is debate over the Saudi capacity a number of sources put it above 12 million b/d (Martyniuk, 1983; Griffen & Teece, 1982). Secondly, there is little doubt that before 1968, much of the Middle East production could have been brought onstream much earlier at very little investment cost, especially in the 'big four'. Indeed so great was the existing known potential that when the possibility of finding reserves in Oman was discussed, Exxon's attitude was that they might put money in it if it was certain 'that we weren't going to get some oil' (Blair, 1977, page 113 quoting Howard Page of Exxon). Similarly, at the end of the sixties there was widespread discussion in the trade press about Saudi capacity of 20 million b/d. For the big four this source of undertstatement therefore more than outweighs any tendency to overstate the figures.

24

How do these estimates of excess capacity compare with other sources? Table 3 contains some estimates from recent sources and when the distortion of different estimates for Saudi Arabia is removed the figures are surprisingly close. Also the Martyniuk and IEA figures are for installed capacity whereas the paper is trying to measure potential capacity. This again suggests the paper's figures are understated.

Comparison problems for the earlier period arise because of a lack of data. As stated much of the literature argues that during the sixties there was limited excess capacity but as explained this referred to excess capacity installed. One source however suggests that at the end of the 1950's excess capacity was of the order of 20-25 percent of capacity and during the mid sixties some 3 million b/d of surplus capacity existed in WOCANA (Seymour, 1980). From Table 2 method two, the excess capacity in 1959 was 23 percent compared to 17 percent in method one from Table 1. Also in 1965 excess capacity is of the order of 3-4.4 million b/d depending on the method. The same source indicates that ARAMCO tried to maintain at least 20 percent spare capacity which compares to a figure of 26.3 percent under method 2 and 18.4 percent under method 1 although this would also include the Neutral Zone.

The next stage is to enquire why this surplus capacity existed? Several explanations are possible none of which is mutually exclusive. The first is that excess capacity conferred on the owner political power and potential influence. This was certainly the case for Saudi Arabia after 1973, especially given its role as swing producer in the 1974-78 period (Stevens, 1982a; El-Mokadem et al, 1983). Secondly, it was an historical hangover from the 1951 Iranian nationalization and other similar supply disruptions such as the Nigerian civil war. These physical supply disruptions meant that for a short period, operational capacity was not available and had to be made up from elsewhere. When the short term disruption was over, government pressure was exerted on the lifters to reintroduce the now available capacity at the expense of the replacement capacity hurriedly developed. Government pressure on occasions could be extremely strong if it was felt production levels were lagging behind historical levels. This pressure was reinforced by the post 'crisis' legacy of high revenue require-ments. For example, witness the pressure from Iran in 1966 (Turner, 1980, page 54 ff.).

This government pressure can be used to refute one of the arguments used by Adelman to assert the absence of excess capacity. He argues (Adelman, 1972, page 164) that if there was excess capacity in 1956-66 of some 20 percent (apart from that required for seasonal and contingency reasons) and Middle East production was growing at 10.5 percent per year then the excess capacity would have disappeared in twenty months by not expanding capacity. Had the companies had a free hand this may well have been so. However, the companies had to keep their host governments at least reasonably happy or face the sort of problems faced in Iraq after 1958. Thus such a run down of excess capacity - apart from Iraq where the dispute relieved some of the capacity pressure - could not be risked without risking the companies' position in the country. For example Howard Page, on relationships with Saudi Arabia, indicated that to ignore the pressure meant they 'could have lost the Aramco concession' (quoted in Blair, 1977, page 108-109). It can be argued that the dispute with Iraq was welcomed by the companies as a relatively easy way of 'removing' some of the governmental pressures. After 1973, this mechanism to create excess capacity was replaced by a different mechanism. The perceived potential supply disruption arising from the Arab-Israeli dispute encouraged companies to develop non Arab oil sources.

The final explanation for the existence of excess capacity which runs throughout the period to the present day concerns the level of crude prices in arms length deals. If the arms length price is held above the replacement cost of proven reserves then the incentive is for companies to develop their own sources of crude supply. Thus to avoid extremes of competition the existing capacity must be held in check even though this effect however unwelcome is to allow the entry of new sources however unwelcome, which may be very high cost. Otherwise strong price competition would follow. This was relevant to 1950 when changes in the US policy towards crude imports effectively meant that new capacity developed specifically to supply the US market was 'stranded' with no down-stream outlets (Hartshorn, 1966, page 42). A similar situation was relevant after the price of crude began to rise in the early seventies. The rent capture by the OPEC countries encouraged other governments to develop crude capacity to allow them to share in the rent bonanza. Put simply, it was a classic supply response to 'high' prices except the increase in quantity supplied was offset by other suppliers restricting supply to prevent a price war.

2.2 Low Marginal Costs

Crude oil production is very capital intensive. Therefore in the short term with capital as the fixed factor, fixed costs are high and variable costs (i.e. marginal costs) are low compared to the netback value of the crude or the arms length price (Adelman, 1972; Bradley, 1967). This is true even in the so called high cost producing areas since the 'high cost' relates to the capital expenditure. In the long term the cost of the marginal barrel is also low in those areas with large proven reserves since the capital cost is being spread over very large output volumes for long periods of time. For example, Adelman presents estimates for the Persian Gulf (Adelman, 1972) of 20 cents p/b and for Africa of 46-54 cents p/b. Even allowing for debate on the exact level of marginal cost, there can be no argument that compared to the crude price in this context it is low.

2.3 Implications of Excess Capacity and Low Marginal Cost

These two characteristics together mean that the pressure exists to increase the sale of crude oil by shading on prices. This needs one qualification. If the seller is a concessionaire then the profitability of extra crude sales will be reduced by the host government's tax take and therefore the pressures to lift more would be muted. If however the seller is a government then no such restraint exists since the tax paid by the national oil company is merely a transfer payment within government. This clearly implies that pressures to increase production by price shading are much more difficult to resist when the seller is a government than a company unless the concession terms to the company are exceedingly generous.

What is the relevance of this to the structure of the international oil industry since 1945? There have been numerous periods when the market has seen downward pressure on prices most recently since 1981. However, an observer of the industry would have expected more pressure and much greater price weakening. It is the hypothesis of this paper (not especially original) that much of the downward pressure on prices has been mitigated by the way in which the industry structure has generally been able

to restrain the excess capacity matching supply with
product demand. The result has been, apart from a
few spectacular changes, relatively steady trends in
the pricing of crude oil - generally downwards. The
pricing history of other mineral commodities has
generally experienced much greater fluctuations.
The remainder of the paper seeks to explain how
integration in the industry has achieved this
situation and the effect when first horizontal and
then vertical integration disappeared or weakened.
It is perhaps worth pointing out at this stage that
this is not simply another academic exercise in
accusing the oil companies of 'hunnish practises'.
Much of the history outlined relates to the
producing governments as well as the companies.

3.   HORIZONTAL INTEGRATION

Horizontal integration is defined as a situation
where different operating units engaged in the same
activity are jointly owned or controlled. This can
occur as a result of merger or agreement. Between
1945 and the early seventies, the horizontal
integration of the industry was embodied in the
joint ownership by the 8 majors of the operating
companies (i.e. Aramco, KOC, IPC and the Iranian
Consortium, etc.) in those areas where the greatest
surplus capacity lay, mainly the big four Middle
Eastern producers. The command over excess capacity
of the eight majors in these 4 countries is given in
Table 4.
     Even outside the big four countries the majors
controlled much of the excess capacity. There were
also some elements of horizontal integration
downstream but these are outside the scope of this
paper.
     This joint ownership arose as a by product of
history. When the concessions were being granted in
the Middle East after the First World War and after
a British company had taken control of Iranian oil,
the United States and French governments were
unwilling to see furtherance of British control.
Thus in both Kuwait and Iraq governmental pressures
secured a presence for their own companies (Penrose
& Penrose, 1979; Sampson, 1975; Stocking, 1970).
Similar pressures also led to the diversification of
the Iranian Consortium as settlement of the 1951
dispute. This diversity of ownership of the
operating companies was reinforced by the operation
of the Red Line agreement (Penrose, 1969; Stock,
1970) which meant that in the area covered by the
defunct Ottoman Empire only joint approaches for

28

concessions would be allowed. The one exception to this was Aramco when the initial concession in 1933 was granted only to one company (Socal). Effectively other companies joined in an effort to find downstream outlets for what was obviously a huge potential crude supply (Penrose, 1969).

This horizontal integration controlled crude supply in one of two ways. The first was by means of a series of checks and balances in the lifting agreements. The APQ system in Iran, the five sevenths rule and its successors in Iraq and the 'dividend squeeze' on overlifters in Saudi Arabia linked to the APQ system (Blair, 1977; Penrose, 1969; Hartshorn, 1966) were all designed to control lifting to match downstream capacities. Furthermore, they were specifically designed to slow down the rate of development of installed capacity as a result either of the time lags involved as in Iraq or as a result of the actual lifting always being below existing capacity as in Iran. Crude short companies were either unable to utilize their equity 'entitlement' to the excess capacity or were very heavily penalized for doing so. This in turn of course encouraged those crude short companies such as Shell to try to develop alternative crude sources which in turn added to the excess capacity problem.

The second way horizontal integration controlled crude supply was by means of the exchange of information inherent in the joint ownership of the operating companies. There was an instant mechanism to detect any company which was seeking to expand its market share at the expense of the others. Everyone knew precisely what everyone else was producing. Textbook economics explain that the uncertainty in oligopoly via the kinked demand curve is a key factor in encouraging stable prices. Companies are unwilling to shade prices because of fear of the unknown, i.e. how their rivals will react. However, in reality this same uncertainty means that if price begins to weaken at the margin the result is a downward helter skelter of prices since no participant can afford to be left behind. On the other hand if a high level of knowledge about rivals' plans is available then price shading is much less likely to lead to an uncontrolled downward spiral. Rather, a sedate decline is more likely to follow unless the 'overlifters' can be controlled as of course they could be by the lifting agreement. For this to occur, overt collusion is unnecessary although of course it can help.

The effects of this supply control were

29

spectacular as can be seen from figures 5 and 6
(following Blair, 1977). Figure 5 maps the
production of 9 of the major world oil exporters.
These are Venezuela, Nigeria, Indonesia, Iran, Iraq,
Kuwait, Saudi Arabia, Abu Dhabi and Libya. The use
of the log scale is intended to emphasise the
changes in production. The dog's dinner which
emerges gives the impression of wild variability
before 1973. After 1973, apart from Iran post
revolution, Iraq post Gulf War and Saudi Arabia
playing swing producer, the others exhibit stable
output tendencies. Yet when the outputs are aggregated in figure 6 the orderly development of crude
production is striking up to 1972. This is not to
make accusations of cartelization against the
companies. The sure symptom of cartelization is
'supernormal profits'. Most evidence shows that
this was not the case for the oil companies and
certainly one study indicates that between 1961-1975
when this system operated, oil company profitability
was no better than other industrial firms (Sunder,
1977). However, these studies are often based on US
oil companies and some question marks have been
raised over the relationship between profitability
and accounting techniques in a US context (Stauffer,
1969). Nonetheless the system certainly stabilized
the profit situation and while profits did decline
in this period (Chase Manhattan) they did not
collapse and nor did any companies go out of
business. Conceivably much of this could be
explained by satisficing behaviour in a potentially
unstable situation.

     The decade of the sixties saw this horizontal
integration come under increasing pressure from two
sources. The relatively high cost of arms length
crude versus the cost of developing owned reserves
encouraged a large number of 'independent' integrated companies to seek new sources of crude and, at
the start of the sixties, Algeria, Libya, Nigeria
and the UAE to name but a few all began oil exports
although the 8 majors were also present in these new
discoveries to varying extents. These new entrants
were hungry for crude. For example, between 1957-
1973 the IRICON agency was top of the APQ
nominations in 9 of the 17 years and second in 5 of
the years (Blair, 1977 Table 5-1).

     Secondly there were new entrants as crude
sellers. The process began with the Soviet Union in
the late fifties and was reinforced when the
American independents were 'stranded' abroad after
1959. The process was further assisted later in the

30

sixties when the National Oil Companies of the
producing countries began to market crude either as
a result of their own operations or by selling
royalty oil to test the market or gain marketing
experience. Inevitably, the erosion of the
horizontally integrated structure was accompanied by
price weakening in both crude and product prices.
By the end of the sixties the realized price of
crude was some 20 percent below the posted price in
most cases (Seymour, 1980).

The system finally appeared to collapse at the
start of the seventies. This was caused by the
change of de jure control which occurred as a result
of a series of nationalizations coupled with
'participation' (Stevens, 1976). As a result the de
jure ownership of the operating companies moved from
the oil majors to the producing governments. This
meant that the lifting agreements (APQ, etc.) became
defunct and with them the access to joint
information. There was now no method to detect
'overlifting' and no mechanism to control overlifting even if it could be detected.

The question must now be asked what kept the
excess capacity in check in the absence of the
horizontally integrated structure? The first point
to make is that at least in terms of installed
capacity there was very much less excess capacity in
existence as is apparent from Tables 1 and 2. Two
factors explain this. The first is the rapid shift
to the right of the demand curve in the late sixties
and early seventies as a result of the economic boom
which began to emerge coupled with a movement down
the demand curve as a result of the real erosion of
oil prices vis a vis other energy sources. This
price fall had been assisted by the companies
increasing production sharply in anticipation of the
change of de jure control already referred to.

Secondly the supply curve shifted to the left
and also changed shape. The reasons for this have
been well documented and can be briefly categorized
into three sections. The first derives from
resource depletion theory in terms of a change in
the discount rate resulting from ownership changes
(Johany, 1979) and changes in price expectations
(Robinson, 1984). The second concerns the
introduction of the Arab oil embargo which involved
a supply reduction in an attempt to side step the
problem of embargoing a country when locating
international crude movements for the countries was
virtually impossible. This factor was only
operative for a relatively short time period.

31

Finally, as the price rose the oil producers supply curve began to bend backwards. Initially this was because they simply did not have the administrative or institutional structure to cope with the massive inflow of revenues (except Kuwait) plus the fact that some had little desire to accumulate foreign assets. Later, as the framework for spending money began to emerge, only the low absorbers who were averse to foreign assets retained this backward bend to the supply curve.

However, this limitation on excess capacity was relatively shortlived as demand fell back following the 1973 price rise and other crude sources responded to the increased profitability by appearing in the market. Thus between 1974 and 1978 two other elements controlled the excess capacity. The first was that the companies continued to act as lifters of the crude both for their own integrated operations and for arms length sales. The reasons for this will be discussed below under vertical integration. The effect was that crude was only lifted when there was a downstream use for it. In other words crude was by and large only moving to end users rather than for speculative sale. Thus the companies would only lift what they required to match crude supply to downstream requirements. In effect there was much the same system as operated before the seventies although probably with less joint information. Fringe speculative sales of course remained in the market but these were relatively small. Towards the end of the period the spot market accounted for only 1-2 percent of the world crude trade (Mohnfeld, 1982). The second element controlling the excess supply was the fact that Saudi Arabia for its own complex reasons (Stevens, 1982a) acted as the swing producer to stabilize the market after it had come to terms with the December 1973 price hike following March 1975. In this role the Saudis relied heavily on the advice of the companies as to what the market required (El Mokadem et al., 1984). Its large range of productive capacity in terms of minimum and maximum production together with its willingness to accumulate foreign assets made it ideal for the role.

The result was that the passing of the old horizontally integrated structure went almost unnoticed in a situation where the sellers of the crude (the countries) and the buyers of the crude (the companies) were both well pleased with the higher prices since both gained and the consumer was able to pay although during 1974-78 the trend of downward pressure on real prices continued.

This alternative to horizontal integration collapsed with the Iranian 'crisis' of 1978-79. The loss of Iranian production was to a considerable extent made up by using excess capacity elsewhere, noticeably from the Gulf. The IOI emerged from the crisis with the old structure in tatters. Most significantly as will be explained below the vertically integrated structure was destroyed and the whole format of contracts had changed from long term preferential contracts to short term contracts and spot sales (Mitchell, 1982; Robinson, 1982; Mohnfeld, 1982; Jensen, 1982; Hartshorn, 1980).

In this context, the spectre of excess capacity reemerged aggravated by the rapidly declining demand for oil although how far this demand decline was due to recession and how far to 'conservation' remains a matter for debate (Stevens, 1982b; Depraires, 1983; Research Group of Petroleum Exporters' Policies, 1982). Immediately the downward pressure on price began. The Saudis tried to reassert their role as swing producer but rapidly reached their lower limit constrained both by revenue requirements, the need for associated gas but most importantly by the requirement that they retained their swing producer role for political reasons (El Mokadem et al, 1984). Also other crude suppliers appeared with a vengeance and in the best traditions of new entrants to an oligopolistic market shaded prices to gain a market share. Even with the loss of Iranian and Iraqi production following the outbreak of the Gulf War in 1980 the excess capacity grew rapidly to levels never dreamt of, even during the 1960's, as can be seen from the earlier data.

Two elements with respect to horizontal integration were relevant. First there was no horizontal integration in any shape or form which meant there was no controlling mechanism on the excess capacity nor was there much knowledge about what rival producers were up to. Furthermore, for reasons discussed below, the refining companies had a vested interest in a lower oil price since for the first time for many of them the crude price to the refinery became a real price rather than an accounting device.

The result was inevitable. Downward price pressure accelerated at a rate which caused concern among the sellers. Some mechanism had to be developed to replace the horizontally integrated

33

structure. In the circumstances only one possibility existed and that was cartelization with production control by OPEC. This was attempted for the first time since 1965-66 in March 1982. The result lasted some six weeks before collapsing in disarray. It was tried again in March 1983 with an output ceiling of 17.5 million b/d coupled with a $5 reduction in price. This appears to be holding although the agreement was and is very precarious, of which more later.

4. VERTICAL INTEGRATION

There exists a vast amount of literature and debate upon why the oil industry developed its vertically integrated structure. There can be said to be two schools of thought on the motives behind the development of vertical integration although they are by no means mutually exclusive. The first argues that vertical integration developed because of the economic advantages inherent in a vertically integrated structure in terms of transactions cost (Coase & Williamson, 1971) and informational advantage (Adelman, 1955). The second school of thought argues that vertical integration was seen as a mechanism for restraint of competition (Blair, 1977; Teece, 1976). To a large extent, the debate has been limited by the fact that it is only in the context of the United States that there has been much empirical attempt to validate the arguments. However, this has tended to fall foul of the fact that the industry in the USA is not particularly vertically integrated. Most studies have shown relatively low concentration ratios (Mancke, 1976; Mitchell, 1976; Ramsey, 1981) although concentration ratios do miss the effects of interlocking directorships (Wilson, 1975).

As to the effects of vertical integration on competitiveness and consumers' welfare again there is considerable theoretical and empirical literature (Machlup & Taber, 1960; Penrose, 1969; Adelman, 1972; Warren Boulton, 1974; Blair, 1977; Department of the Treasury, 1976; Greenhut & Ohta, 1979; Carlton, 1979; Beckenstein et al., 1979; Waterson, 1982). However, time and space do not allow for further discussion of this literature. In WOCANA, the oil industry certainly has had a very high degree of vertical integration between 1945 and the early seventies (Penrose, 1969; Adelman, 1972; Blair, 1977). Rather than discuss the impact of vertical integration on general industry competitiveness it is intended to concentrate on the

impact of the vertically integrated structure on the
crude oil market in the context of excess capacity
and low marginal cost partly because this is the
issue of the paper but also because vertical
integration has received extensive although
incomplete attention.

The existence of vertical integration clearly
encouraged the control of the excess capacity. To
have lifted crude above the downstream capacity of
the company would have meant selling at arms
length. While this would generate profits, it would
also generate greater competition in the product
markets. In so far as this meant a loss of market
share then less 'integrated' crude would be required
and hence less profit made on that crude.
Downstream refining was in any case unprofitable
because of transfer pricing policies. Thus crude
lifters in this period, namely the eight majors,
would sell only to end users - either themselves or
other refiners. The question may then be asked, why
sell to other refiners? The answer is that if they
did not these other refiners would have an incentive
to seek alternative crude supplies aggravating the
over supply situation (Penrose, 1965). Linked to
this is the fact that 'the integrated firm has
chosen to take on a heavier burden of fixed costs to
gain lower average costs of production or greater
certainty. He loses flexibility in the process ...
(and) therefore has a natural preference for
stability in the market place' (Caves, 1977, page
44). This preference for stability would reinforce
the tendency to restrict crude lifting to meet the
integrated needs of the companies together with
sufficient crude lifted for arms length sale to
minimize the incentive to develop crude elsewhere.
It can be argued that in fact the oil majors
misjudged this balance. By allowing too little arms
length crude into the market they encouraged the
development of capacity elsewhere. Thus up to the
early seventies, the vertically integrated structure
reinforced control over the excess crude producing
capacity.

With the take over of de jure control by the
producing countries in the early seventies it was
generally expected that the vertically integrated
structure would disappear with the loss of supplies
of 'owned' crude. In fact this did not happen. The
major companies continued to lift and dispose of the
crude oil on behalf of the governments. This
included third party sales and in 1973 some 22 percent of the crude available to the seven majors went

to third party sales while in 1978 the figure was still around 16 percent (Mohnfeld, 1982).

Several explanations are possible to explain why the countries allowed the companies to retain this preferential access to crude (Hartshorn, 1980). They revolve around the lack of crude marketing expertise in the countries coupled with a weakening crude market after 1974. There is a further explanation related to the nature of government administration in many of the OPEC countries. Oil trading is highly complex. The result is that it is very difficult to assess whether or not a crude sale has been at a 'good' or 'bad' price even for those expert in the field. After the 1973 price rise, many of the governments became extremely sensitive about the issue of bribery and corruption in contracts of all types. In such a context, it is the brave official who would sell crude - especially in a buyers market - and risk subsequent accusations of selling the crude at too low a price for 'special favours'.

This situation changed dramatically with the Iranian crisis of 1978-79. The initial feature of the crisis was that spot crude prices determined in the market place rose faster than government official selling prices which were administratively determined (El Mokadem et al., 1984). This meant that companies with access to preferential crude on long term contracts were able to make a 'killing' in the market. Not unnaturally, the governments saw this situation as unacceptable and therefore began to abrogate the long term contracts pleading force majeure and began to market their crude directly. Earlier fears of obtaining poor prices disappeared in the strong sellers market generated by the crisis. The result was that the companies lost their marketing role for crude (Torrens, 1980; Mitchell, 1982; Mohnfeld, 1982; Robinson, 1982). By 1982, private oil companies held title to less than 20 percent of production and had access to only a third of production on long term preferential contract compared to over 90 percent in 1970 (hartshorn, 1983). Effectively, the second oil shock destroyed much of the vertically integrated structure in WOCANA.

Three consequences followed from this change in structure. The companies for the first time had a growing interest in lower crude prices, the crude oil market became much more competitive (at least prior to the so far successful cartelization of March 1983) and finally there was a dramatic loss of

36

joint information within the industry.

The loss of preferential access to crude also meant the loss of profits attributable to that access. The 'profit centres' for the companies to a great extent had become downstream centres. This is not to say that owned crude was not still profitable. One study for 1981 suggests that oil company profitability on sales was 18.1 percent on chemicals and 6.7 percent on coal (ENI, 1983). However, the profits from production of crude and gas were now a much smaller base from which to calculate total profit. This 'loss' of profit had two effects. The first was a scramble to find alternative sources of owned crude. This was achieved either by investment in exploration and development or more recently an attempt to buy 'off the shelf' proven reserves by company take over and merger. In 1980, some 66 percent of oil company investment went on exploration and production compared to 40 percent in 1970 and 41.5 percent in 1975 (Chase Manhattan). Also the financial press has been full of details of company take overs and the relative costs of acquiring reserves by investment or purchase. The second effect of the 'loss' of profit was that each stage of the industry had to become profitable in its own right (Algar, 1983). Thus the refining sections would have an interest in lower crude prices since lower input costs clearly increase refinery margins, other things being equal.

These two elements have lead to an almost schizophrenic dichotomy for the oil companies. Lower prices mean higher refinery margins but lower margins on owned crude production. The balance of interest lies in the absolute magnitudes of both profit centres coupled with the vulnerability of the before tax profits in both areas to government tax policy.

The second consequence of the collapse of the vertically integrated structure is that the crude market has become more competitive. The majority of oil now moves either on short term contracts or in the spot market. It has been suggested that some majors are using the spot market for up to 50 percent of their crude requirements (Algar, 1983) although to this author that figure seems far too high. This is partly because in most cases long term contracts became totally discredited during the Iranian crisis. It is also the consequence of being a buyer in a buyer's market. It now pays to 'shop around' and this is not possible when constrained by long term contracts. In such a context, surplus

producing capacity required drastic action to
prevent a major price collapse. This appeared
eventually in the form of the London Agreement in
March 1983 (El Mokadem et al. 1984). However, it is
worth pointing out that the extent of the surplus
capacity is now so great that without a major supply
disruption, the cartel will have to hold for a very
long time if price is not to collapse at some point
in the future. The strains on the cartel from
revenue and associated gas needs are such that there
seems only an outside chance that the cartel will
hold.

The first consequence of structural change is
the loss of joint information. With neither
horizontal nor vertical integration no one knows
what rivals are up to. This is clearest within OPEC
when the official Ministerial Monitoring Committee
is forced to publish several estimates of members'
production. A good example of the consequences came
in the summer of 1983. Early in the summer, demand
for crude in Europe picked up possibly due to fears
for supply security on the announcement that the
French would supply the Etendard/Exocet combination
to Iraq. The result was that for the first time
since the London Agreement spot crude prices
reached, and in some cases exceeded, official
prices. Almost immediately most of the Gulf
producers decided to increase spot sales to take
advantage of the new situation. This was reflected
in a sharp rise in the spot tanker rate. However,
the result was a classic example of the fallacy of
composition. Because everyone had independently
read the same signals and reached the same decision,
the spot price fell back rapidly as these new
supplies came into the market.

It has been argued that the loss of
information in this area has been offset by greater
transparency on pricing. It is certainly true that
the increase in the volume entering the spot market
has made the spot price movement more meaningful
than before. Some commentators also suggest that
the development of futures markets may well achieve
the same result (NYMEX, 1982) although this author
has very strong reservations on that point. However, given the time lags inherent in the oil
trading business all the price signals tell the
market is what decisions were made some time ago.
This is not always a good guide for decision taking
now which will receive no feedback until some time
in the future. If this sounds rather like a
guessing game this is precisely what the oil market

has become since the vertically integrated structure
went the way of the horizontally integrated
structure. With the existence of so much spare
crude capacity, this makes the oil market akin to a
situation of asking a bright six year old to defuse
a bomb acting on instructions shouted from a
distance. Either the wind may blow away whether it
is the blue or red wire to cut or the child gets
bored and decides to kick the bomb instead. As the
boredom option becomes more likely so the distance
from which the instructions are shouted grows.

6.   CONCLUSIONS
Two broad conclusions suggest themselves. The first
is that the IOI exhibits remarkable continuity in
terms of the problems which have been created by
excess capacity over time. What has changed has
been the method of containing the excess capacity
and the relative success achieved by the different
methods in controlling supply to avoid excessive
price competition in the crude oil market. However,
a further continuity is that all the methods
employed have shown a tendency eventually to 'fray'
round the edges. In particular it appears that
government 'solutions' are more vulnerable than
'company' solutions because the rewards on ignoring
the supply constraints are greater for governments
who also capture the tax element of the rent. The
second broad conclusion is that industry structure
is shaped either as a result of economic forces such
as transaction costs and informational advantage
linked with historical accident or it is shaped by a
deliberate attempt on the part of the participants
to contain competition. The latter I call a
'contrived' structure. At times it appears both
elements have been at work simultaneously. In
either case the result is restricted competition.
However, it may be that without some causal element
which goes beyond simple competition restriction
i.e. causes other than a 'contrived' structure then
the structure 'frays' around the edges more quickly.

READING LIST

ADELMAN, M.A. (1955) Concept and Statistical Measure of Vertical Integration. In NBER, Business Concentration and Public Policy, Princeton University Press.

ADELMAN, M.A. (1972) The World Petroleum Market, John Hopkins University Press.

ALGAR, P. (1983) Oil Majors - Changing Times and Pulling in the Reins, Energy Economist, Nos 24 and 25.

BECKENSTEIN, A.R.; GRAYSON, L.E.; OVERHOLT, S.H. and SUTHERLAND, T.F. (1979) Performance and Measurement of the Petroleum Industry, Lexington Books.

BLAIR, J.M. (1977) The Control of Oil, Macmillan.

BRITISH PETROLEUM. Statistical Review of the World Oil Industry. Various Years.

BRADLEY, P.G. (1967) The Economics of Crude Petroleum Production, North Holland.

CARLTON, D.W. (1979) Vertical Integration in Competitive Markets Under Uncertainty. Journal of Industrial Economics, March.

CAVES, R.E. (1977) American Industry: Structure, Conduct and Performance, Prentice Hall.

CHASE MANHATTAN, (Various Years) Financial Analysis of a Group of Petroleum Countries.

COASE, R. and WILLIAMSON, O. (1971) Vertical Integration of Production. Market Failure Considerations. American Economic Review 1971.

CREMER, J. and SALEHI-ISFAHANI, D. (1981) Competitive Prices in the Oil Market, Mimeographed Paper.

DEPARTMENT OF THE TREASURY (1976) Implications of Divestiture, Department of the Treasury, USA.

DESPRAIRES, P. (1983) Short and Long Term Prospects for the Energy Market. SEED Paper No 13, February, Surrey Energy Economics Centre.

EL-MOKADEM, A.M.; HAWDON, D.; ROBINSON, C.; STEVENS, P.J. (1984) OPEC and the World Oil Market 1973-1983, Eastlords Publishing.

ENI, (1983) Strategies of Oil Companies, Mimeographed Paper.

GREENHUT, M.L. and OHTA H. (1979) Vertical Integration of Successive Oligopolists, American Economic Review, March.

GRIFFEN, J.M. and TEECE, D.J. (1982) OPEC Behaviour and World Oil Prices, George Allen and Unwin.

HARTSHORN, J.E. (1966) Oil Companies and Governments, Faber and Faber.

HARTSHORN, J.E. (1980) From Multinational to

National Oil: The Structural Change. Journal of Energy and Development, Spring.

HARTSHORN, J.E. (1983) Re-Cast Roles in the World Oil Market. Middle East Economic Survey (MEES) Vol XXVII No 2.

HOTELLING, H. (1931) The Economics of Exhaustible Resources, Journal of Political Economy Vol 39 April.

IEA, (1976) World Energy Outlook, IEA.

JENSEN, J. (1982) New Territory in the World Petroleum Market. MEES Vol XXV, 14.

JOHANY, A. (1979) OPEC and the Price of Oil: Cartelization or Alteration of Property Rights, Journal of Energy and Development, Autumn.

MACHLUP, F. and TABER, M. (1960) Bilateral Monopoly, Successive Monopoly and Vertical Integration, Economica, Vol 27.

MANCKE, R.B. (1976) Competition in the Oil Industry, in Mitchell 1976.

MARTYNIUK, R.J. (1983) Past, Present and Future Trends in Oil Reserves and OPEC Production Capacity, Energy Exploration and Exploitation, Vol 1 No 4.

MITCHELL, E.J. (1976) Vertical Integration in the Oil Industry, American Enterprise Institute for Public Policy Research (AEI).

MITCHELL, J. (1982) Anatomy of an Oil Crisis, Energie Wirtschaft, June.

MOHNFELD, J.H. (1982) Implications of Structural Change, Petroleum Economist, July.

NYMEX, (1982) Energy Futures. New York Mercantile Exchange.

PENROSE, E.T. (1965) Vertical Integration with Joint Control of Raw Material Production: Crude Oil in the Middle East, Journal of Development Studies, Vol 1 No 3.

PENROSE, E.T. (1969) The Large International Firm in Developing Countries, George Allen and Unwin.

PENROSE, E.T. and PENROSE, E.F. (1978) Iraq, Ernest Benn, Westview Press.

RAMSEY, J.B. (1981) The Oil Market: Control versus Competition, Ethics and Public Policy Centre.

RESEARCH GROUP OF PETROLEUM EXPORTERS' POLICIES, (1982) Oil Prices in 1983: A Critical Year. MEES XXVI No 8.

ROBINSON, C. (1984) The Changing Energy Market: What Can We Learn from the Last Ten Years? in Hawdon, D. (Ed.) The Energy Crisis Ten Years After, Croom Helm.

ROBINSON, M.S. (1982) The Crude Oil Price Spiral of 1978-1980, Mimeographed Paper.

SAMPSON, A. (1975)   The Seven Sisters, Hodder and
    Stoughton.
SEYMOUR, I. (1980)   OPEC Instrument of Change,
    Macmillan.
STAUFFER, T.R. (1969)   Measuring the Profitability
    of Petroleum Operations.   Paper to the 8th Arab
    Petroleum Congress.
STEVENS, P.J. (1976)   Joint Ventures in Middle East
    Oil 1957-1975, M.E.E.C.
STEVENS, P.J. (1982a)   Saudi Arabian Oil Policy in
    the 1970's - Its Origins, Implementations and
    Implications.   In Niblock, T. State, Society and
    Economy in Saudi Arabia, Croom Helm.
STEVENS, P.J. (1982b)   The Future of World Oil
    Prices: The End of an Era?, ODI Review, No 2
    1982.
STOCKING, G.W. (1970)   Middle East Oil, Vanderbilt
    University Press.
SUNDER, S. (1977)   Oil Industry Profits, AEI.
TEECE, D.J. (1976)   Vertical Integration in the US
    Oil Industry, in Mitchell, 1976.
TURNER, L. (1980)   Oil Companies in The
    International System, George Allen and Unwin.
TORRENS, I.M. (1980)   Changing Structure in the
    World Oil Market, The Atlantic Institute for
    International Affairs.
WARREN BOULTON, F. (1974)   Vertical Control with
    Variable Proportions, Journal of Political
    Economy, Vol 82.
WATERSON, M. (1982)   Vertical Integration, Variable
    Proportions and Oligopoly.   Economic Journal,
    March.
WILDAVSKY, A. and TENENBAUM, E. (1981)   The Politics
    of Mistrust: Estimating American Oil and Gas
    Resources, Sage Publications.
WILSON, J.W. (1975)   The Market Structure and
    Interfirm Integration in Oil, Journal of
    Economic Issues, June.

A Survey of Structural Change

TABLE 1

TOTAL EXCESS CAPACITY IN WOCANA BY PERCENTAGE OF TOTAL CAPACITY
AND PERCENTAGE OF ORIGIN OF EXCESS CAPACITY BY REGION

METHOD 1 PRE 1968 2 YEAR LEAD POST 1968 1 YEAR LEAD

YEAR	TOTAL	WEST HEMIS	AFRICA %	EUROPE	BIG 4 MID EAST	OTHER MID EAST	OTHER EAST HEMIS
1959	17	21	19	3	53	2	2
1960	16	25	26	3	40	2	3
1961	17	22	28	2	42	4	2
1962	17	12	33	3	46	5	1
1963	17	11	31	2	46	8	2
1964	15	4	36	1	52	4	4
1965	15	7	33	1	48	7	4
1966	17	12	41	0	39	3	5
1967	19	5	41	0	39	8	7
1968	17	4	34	0	45	7	11
1969	8	7	23	0	58	0	12
1970	8	1	0	0	88	0	10
1971	9	5	27	1	60	0	7
1972	14	7	22	0	62	0	8
1973	7	2	46	0	50	0	1
1974	8	4	58	5	31	0	1
1975	17	1	35	4	55	0	5
1976	11	5	38	11	40	0	6
1977	10	10	35	10	42	2	0
1978	16	8	25	7	55	3	2
1979	14	9	19	3	66	2	1
1980	22	6	17	3	70	3	2
1981	31	5	21	2	65	5	1
1982	38	7	18	0	67	6	1

SOURCE: Author's estimates based on British Petroleum data
(British Petroleum)

TABLE 2

TOTAL EXCESS CAPACITY IN WOCANA BY PERCENTAGE OF TOTAL CAPACITY
AND PERCENTAGE OF ORIGIN OF EXCESS CAPACITY BY REGION

METHOD 2 PRE 1968 3 YEAR LEAD POST 1968 2 YEAR LEAD

YEAR	TOTAL	WEST HEMIS	AFRICA %	EUROPE	BIG 4 MID EAST	OTHER MID EAST	OTHER EAST HEMIS
1959	23	25	15	3	50	3	4
1960	25	24	22	3	47	3	2
1961	23	21	28	2	42	5	2
1962	25	17	30	2	43	6	2
1963	24	11	32	2	45	8	1
1964	24	7	34	1	47	7	3
1965	22	7	34	1	47	7	4
1966	24	8	40	0	42	6	4
1967	24	8	37	0	40	8	7
1968	27	5	35	0	42	10	8
1969	17	4	34	0	45	7	11
1970	17	4	20	0	63	4	10
1971	16	0	9	0	75	7	9
1972	19	2	16	0	69	3	9
1973	17	6	20	0	62	5	7
1974	8	-7	46	6	49	5	1
1975	9	-14	54	13	39	0	8
1976	21	3	29	9	48	4	7
1977	14	11	33	16	35	0	5
1978	15	13	24	14	45	2	3
1979	19	13	21	9	52	2	2
1980	15	15	18	5	61	1	0
1981	23	10	16	5	65	3	1
1982	32	9	20	2	62	5	1

SOURCE: Author's estimates based on British Petroleum data
(British Petroleum)

TABLE 3

COMPARISONS OF CAPACITY FROM SELECTED SOURCES

	1976 Tables 1 & 2	1976 IEA	1981 Tables 1 & 2	1981 Martyniuk
Venezuela	-	-	3.2	2.6
Nigeria	-	-	2.3	2.5
Libya	3.2	2.5	3.2	2.5
Iran	6.0	6.5	6.0	6.0
Iraq	2.4	3.0	3.4	4.0
Kuwait	3.0	3.5	3.0	3.0
Indonesia	-	-	1.6	1.8
TOTAL	14.6	15.5	22.7	22.4
Saudi Arabia	9.1	11.5	9.9	12.5

SOURCES: Martyniuk, 1983; IEA, 1976

TABLE 4

AVERAGE EQUITY ACCESS TO EXCESS CAPACITY 1958-1970 MILLION
TONNES PER YEAR

	BP	SHELL	CFP	EXXON	MOBIL	GULF	TEXACO	SOCAL
Iran	22.52	7.88	3.38	3.94	3.94	3.94	3.94	3.94
Iraq	4.84	4.84	4.84	2.42	2.42	-	-	-
Saudi	-	-	-	12.8	4.3	-	12.8	12.8
Kuwait	14.7	-	-	-	-	14.7	-	-
TOTAL	42.06	12.72	8.22	19.16	10.66	18.64	16.74	16.74

SOURCE: CHASE MANHATTAN

45

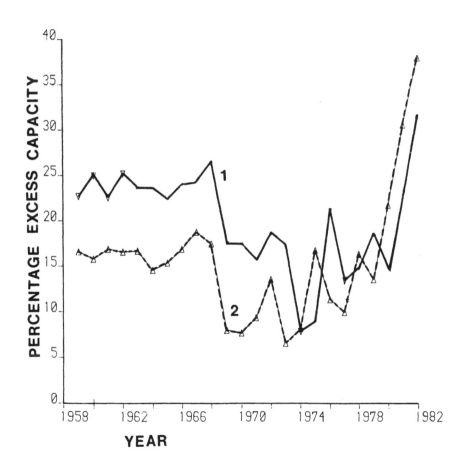

FIGURE 1

EXCESS CRUDE PRODUCING CAPACITY
WOCANA — METHODS 1 AND 2

PERCENTAGE OF TOTAL CAPACITY

FIGURE 2

REGIONAL ORIGIN OF WOCANA EXCESS CAPACITY — METHOD 1

PERCENTAGE

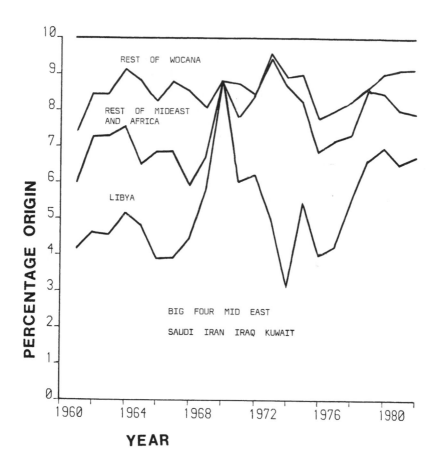

47

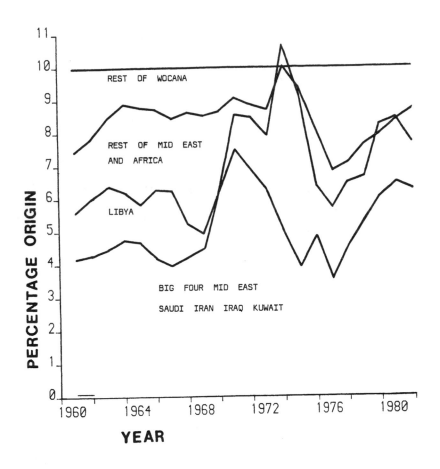

FIGURE 3
REGIONAL ORIGIN OF WOCANA EXCESS
CAPACITY – METHOD 2

PERCENTAGE

REST OF WOCANA

REST OF MID EAST
AND AFRICA

LIBYA

BIG FOUR MID EAST

SAUDI IRAN IRAQ KUWAIT

PERCENTAGE ORIGIN

YEAR

FIGURE 4
TOTAL POTENTIAL CAPACITY AND PRODUCTION — WOCANA
MILLION B/D

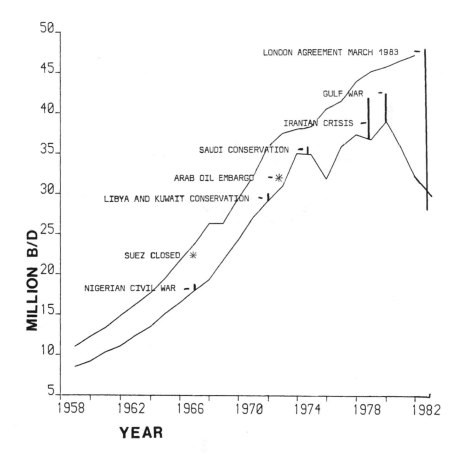

* INDICATES INDETERMINATE 'LOSS' OF CAPACITY

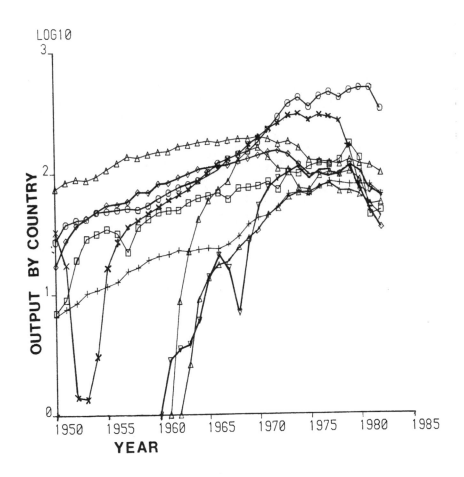

FIGURE 5
CRUDE OIL OUTPUT
NINE MAJOR WOCANA PRODUCERS
MILLION TONNES

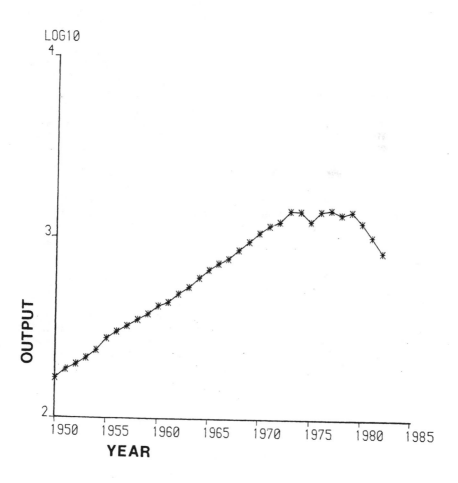

FIGURE 6
TOTAL CRUDE OIL OUTPUT
BY THE NINE WOCANA COUNTRIES
MILLION TONNES

3 STANDING THE TEST OF TIME

Richard G Reid

It seems a particularly appropriate moment to
comment on the oil industry's recent past and
reflect on its future. Ten years ago, almost to the
day, the Arab members of OPEC lifted the oil embargo
that they had imposed on October 17, 1973. For
Exxon and all the major oil companies, those five
months marked a turning point in our evolution. We,
along with the rest of the world, are still
adjusting to the changes in the oil prices, demand
and supply that took place during the oil embargo,
not to mention the jolts we experienced during the
Iranian revolution.

Until the oil embargo, oil prices were
essentially constant through the 1960's and
increased only slightly in the early 1970's. At
that time we thought we saw some fluctuations in the
market, but we had no idea of what was to come.

For example, in 1971, Exxon forecast that over
the next ten years, the price of Arabian light crude
would not rise above $3 a barrel. Then in
1973-1974, the price of oil suddenly quadrupled. By
the time 1981 rolled around, the oil market had
experienced another price explosion and the official
price of Arabian light crude had climbed to the
level of $34 a barrel.

Looking back on our forecasts, we had only the
slightest notion of how much the price of oil would
rise and how much the change in prices would affect
oil consumption over the long term. In 1974, Exxon
projected that over the next ten years oil demand
worldwide would grow at an average rate of 3 percent
a year instead of the 7-8 percent annual growth
experienced in the previous decade. It was
impossible for us to imagine that non-Communist
world oil demand in 1983 actually would be 6 percent
less than it was in 1973 and about 40 percent below

52

the level for 1983 that was estimated in the 1974 forecast.

Changes in pricing and demand coincided with changes in the commercial access to oil. Until the embargo, the multinational oil companies had the predominant equity interests in the oil production of most OPEC member states. Prior to the 1970s companies such as Exxon made the important decisions on the level of investments and the rates of production. In addition, these companies maintained integrated logistics and marketing systems that provided an outlet for their production. Finally, the integrated majors were the main suppliers of crude oil to independent refiners.

Before I go any further let me define some jargon that we in the oil industry constantly use. By upstream we mean exploration for and production of oil and gas and by downstream we mean refining, transportation and marketing of oil.

Over the last ten years the majors have lost their equity producing interests in most OPEC countries. Whereas in 1973, about 80% of crude oil production was from privately owned companies such as Exxon, by 1982 their share had dropped to a little over 10 percent. In the same period, production from national oil companies grew from a share of around 20 to 90 percent. Although Exxon still has important operating arrangements in Saudi Arabia, Venezuela and Indonesia, these terms of access are not the same as the old equity crude rights.

These striking changes in crude oil ownership, demand and oil prices have changed the size and shape of our business. In the early 1970s Exxon's equity crude production was about 6 mb/d, about 75 percent of which was located in the OPEC countries. Today we produce 1.6 mb/d, only about 3 percent of which is in OPEC countries. In the early 1970s we sold 6 mb/d of oil products and had confidence in our forecasts of steady growth. Today we are essentially a 4 mb/d company with the prospect of very little sales growth. Finally, in the early 1970s, the level of oil prices was viewed to a certain extent as a constant in a decision-making process that contained several other variables. Today, the price of oil is the chief variable in our equations and the greatest single source of uncertainty about the future.

In view of these realities, Exxon like many of the major oil companies has had to alter the basis on which it runs its business. There are three main

53

points I would like to make in characterising our
strategy today.

First, our top priority is to acquire
promising exploration acreage and to delineate and
develop new reserves in order to safeguard the
continuity of oil and gas supplies in the years
ahead.

Second, each segment of our business must
prove its individual worth in the marketplace.

And third, we are well aware of the oil
market's imprecisions and volatility, as well as our
limited ability to forecast accurately its
movements. Nevertheless, we still believe that the
market place is our best guide for investment
decisions.

Let me elaborate on each of these points.

Over the past ten years, more oil has been
produced than discovered with the result that the
world's inventory of discovered reserves has
gradually declined. We estimate that in 1982, new
oil discoveries worldwide were running at only about
two-thirds of the level of oil production and we
expect that trend to continue. That poses a
considerable challenge to all oil producers but
especially those outside the OPEC countries.
Non-OPEC oil is being produced close to full
capacity at about 25 mb/d and is likely to hold more
or less steady at this volume to the year 2000. But
for this to be true, nearly 40 percent of supply in
the year 2000 will have to come from deposits not
yet found. So, during the next 17 years around 50
billion barrels of non-OPEC oil will have to be
discovered.

Finding and developing those reserves will be
progressively more difficult and costly. Over the
years, only one out of ten exploratory wells drilled
worldwide has proved to be successful. Furthermore,
many of the fields that are being found are smaller
and have a more complex geological structure than
the large reservoirs that were previously
discovered. For example, the largest field ever
found in the North Sea contained just under 3
billion barrels while most of the new fields contain
less than 100 million barrels.

In line with this worldwide trend, Exxon has
also had to intensify its search for new oil and
gas. Over the past ten years, our worldwide proved
reserves of crude oil have shrunk from 38 billion
barrels in 1973 to 6.5 billion barrels at the end of
1983 due largely to the loss of ownership in OPEC
countries as well as the depletion of reserves in

North America. Today our principal concern is to reverse that decline. Reserves increased slightly in 1983 following a 3% increase in 1982. My company is searching for oil and gas wherever we think there is an attractive opportunity offering a good return on our investment. We are drilling wells all over the world, from China to Mozambique, from well above the Arctic Circle to Australia. Last year, around the world, Exxon had a total capital investment programme of over $9 billion. Of this, about $6.1 billion was spent exploring for and developing crude oil and natural gas resources and we plan to spend considerably more in 1984.

Incidentally, the upstream commands an increasing portion of our capital and exploration expenditures. In 1973, spending on exploration and production represented 46 percent of Exxon's total expenditures. By 1983, they accounted for 67 percent. Furthermore, we have been realising a larger percent of our total earnings in the upstream. Earnings from petroleum and natural gas production as a percent of our net income rose from 72 percent in 1973 to 82 percent in 1983. So, while the events of the past decade have raised the capital requirements and the risks of exploration and production, the rewards resulting from our upstream activity have also increased.

Since we expect that the greater part of our investment and earnings will remain upstream we have had to reassess our approach to other parts of our business. This brings me to my second point.

For many years Exxon has taken the view that each segment of its business should be economically justified and stand on its own feet in the marketplace. The facts of the past ten years have left us little choice but to ensure that each activity in every geographic part of our business meets this test.

Those of you who attended this conference last year heard Sir Peter Baxendell, the Chairman of Shell on this subject. He pointed out that the days were gone when losses in some areas could be balanced by profits made in others. Let me quote what he said:

"There is a need for each part of the business to be viable in its own right, and each should indeed be treated as a separate business area, with the links between forged on the basis of economic and competitive practice".

We are planning to stay in the downstream not because we need it as an outlet for our equity crude but because we believe it will be financially self-supporting. Our equity crude is being produced essentially at full capacity, and still it supplies only about 48 percent of our refinery runs. We would not keep a downstream business of nearly the size that we have unless we believed in its long term profitability in its own right.

Our basic strategy is to make our company at least equal to the most efficient competitor in every market in whcih we participate. So, we are making a determined effort to streamline the refining and distribution network worldwide. But we cannot achieve that goal by simply closing down old facilities. We are currently modernizing and upgrading the system - at considerable cost. For example we recently started construction of a major conversion unit at our Rotterdam refinery that will take the heaviest portion of a barrel of oil and convert it entirely into lighter products. For the project to be profitable, however, the light products will need to have a significantly higher value than the heavy oil used as feedstock. This spread in price will depend on the supply and demand of the various petroleum products as well as coal and natural gas prices. We had to make a forecast of what the differentials might be and then bet on it - a billion dollar bet.

There are some political uncertainties that compound the economic problems posed by our downstream business. One can imagine conditions under which, no matter how efficient we were, we could not hope to be profitable. Consuming country governments could control product prices at levels below costs. It has happened before and could happen again. East bloc and other producing countries seeking outlets for their crude oil through the product markets could price at levels that would make it extremely difficult for even the most efficient companies to be successful.

The final point I wish to make is that despite these political and economic uncertainties, the market remains the best guide for investment decisions. While our hearing like our vision of the future isn't perfect, we listen closely to what the market is saying. We know that it isn't a pure system but we are convinced that by and large the market-place and its pricing signals offer the most reliable guide to what should be done. But it does, almost always, tell us when things are going wrong and give us a chance to get back on course.

Our experience with synthetic fuels serves to

illustrate this point. In the wake of the 1979-1980 price increases, Exxon started a number of synthetic fuel projects including Colony Shale in Colorado, Cold Lake Heavy Oil in Canada, and the huge Oil Shale project at Rundle in Australia. These projects all had long lead times of ten or more years and each would have required the investment of several billion dollars. So we were taking preliminary steps based on our perceptions of the 1990s. In total they were expected to strain the resources of even a company as large as Exxon.

However, due to the changed outlook of reduced demand, lower prices and lower growth, all of these projects had to be deferred or significantly reduced. I stress that we have delayed, not permanently cancelled these ventures. We continue to believe that at some time in the future, synthetic fuels from these sources and from coal not only are going to be required but will offer us a good return on our investment. Therefore we selectively maintain research programmes and resource bases so that we will be prepared to grasp the opportunity when the market tells us the time is right. However, with today's excess supplies of oil and of other forms of energy, we do not expect projects to be started again for some years to come.

What I have just said about synthetic fuels also applies to conventional energy projects. The market can and should play the key role in setting the future pace of resource development - be it coal, petroleum or nuclear energy. We recognise the uncertainties that exist in today's market and the risks that we incur by relying on it. But when we ignore the market, we dig a deeper hole for ourselves.

I feel that many governments have also come to recognise that the less they attempt to manage the market, the better off they and their economies are. For example, not too long ago, most European governments had price controls on oil products. Today, however, only a few governments maintain these controls and their systems are becoming more market oriented each year. There was also a time when direct government-to-government crude oil deals seemed to be the trend of the future. But many governments were discouraged by the numerous complications that accompanied those agreements. As a result, most of those agreements have lapsed and the world's oil trade continues to flow primarily through the channels established by the commercial oil companies.

That brings me to a point that I feel does not receive enough emphasis. The international oil companies are non-political organisations that can provide a buffer between governments during periods when political tensions are running high. We are commercial organisations with a demonstrated ability to adapt to vastly changing circumstances whether they be of a political or economic nature.

But in order to adapt and survive, we have had to maintain the confidence that we can remain profitable in the long run in spite of all the risks and uncertainties. We believe that private companies have a significant contribution to make to society and the economy, but they can only do so if they can make a reasonable return on their investments.

Just as companies keep their ear attuned to the market, so governments need to heed those market signals. Taxation as well as prices determine the viability of any project. Lower taxes could make feasible resource exploitation that is today too costly. Higher taxes could have the opposite effect. We feel that it would be in the interest of governments to moderate oil and gas taxation in a way that is consistent with recent decreases in oil prices and increases in costs.

In this regard we have been greatly encouraged by the approach that the British government has taken in its tax and energy policies and hope that other governments involved in over-seeing petroleum and other energy projects will follow the British example. Incidentally, our response to the new regime has been swift and I hope will be productive. Esso's exploration and appraisal drilling in the North Sea was at record levels in 1983 and will be almost twice as high in 1984. Included in this activity will be the drilling by Esso of a wildcat well west of Shetland in a water depth of over 2000 feet. So if both governments and companies pay attention to the signals that the market is emitting, I feel sure that we can maintain strong petroleum exploration, development and production in the coming years.

I would like to end by saying that few companies in recent years have faced greater change, bigger risk or decision making more momentous for society than the large private oil companies. But despite the vicissitudes of the past decade, they have stood the test of time and will continue to do so.

4 GOVERNMENT SELLERS IN A RE-STRUCTURED CRUDE OIL MARKET

Jack E. Hartshorn

Two aspects of the structural change we are
discussing today have altered international crude
supply radically, but in rather contary ways.
First, it has greatly enlarged the arm's length
crude market - which in most senses is the only real
market for crude oil. Secondly, the main actors in
this enlarged market are all largely new to trading
in oil, and by nature are unlike all other traders.
The first of these developments is tending to open
out the whole market in oil. The second - perhaps
inherently because these new actors are governments,
but in the event as a matter of chosen policy - may
tend, and is certainly intended, to tighten this
market. These two concomitants of the same
structural change interact across today's whole oil
market. That market remains precarious. Price
levels, which are slipping gradually in real terms,
are still high enough to keep the response of demand
to economic recovery only sluggish. At the same
time prices are so far above any current producer's
supply costs that no supplier has the incentive to
drop out.

ENLARGED ARMS LENGTH MARKET
Although the volume of crude moving in inter-
regional trade is about 40 percent lower than a
decade ago, and OPEC exports have fallen by close on
half, arm's length sales of crude have probably
doubled in volume. As a proportion of crude moving
in this trade, their importance may have trebled.

This is no paradox. It is simply a mirror-
image of the declining coverage of vertical
integration in the world movement of crude. When
integrated supply through the former seven international majors' channels was the dominant mode of
world oil supply, the volume of crude oil sold at

arm's length was in fact rather small.

Most of the crude which then moved internationally was never sold at all. It moved from
equity producers to affiliated refiners at inter-
corporate transfer prices. Economists would not
accept those prices as constituting market prices.
What is more important, neither would many fiscal
authorities. It did not change ultimate beneficial
ownership until, refined into products, it reached
consumer markets downstream - the one place where
all oil is, finally, sold.

Until the seventies, third party transactions
in crude probably didn't exceed 20 percent of the
total volumes moving. Since contracts then often
had a three-year term, it used to be estimated that
only about 7 percent of annual crude supply was
priced in arm's length deals made in any year.
Major company spokesmen offering such rationales
emphasised how narrow a market that was, and that
prices realised in it, sometimes resulting from
distress sales, could not be regarded as
representative. (To non-economists, they often
described such third-party sales as "quite
marginal". To economists, they were at pains to
argue that marginal was just what such prices were
not). You can compare that with the views expressed
by established companies only a few years ago, and
by some OPEC spokesmen right up to now, about the
spot market. The adjectives, and the attitudes,
were just the same.

By 1972, just before general participation in
the Gulf, third party sales of crude were perhaps
6 - 8 million barrels a day, out of nearly 25
million barrels a day moving inter-regionally. But
around 6 million barrels a day of the third party
sales were made by the integrated majors themselves
who were at that time producing 22 million barrels a
day, almost entirely equity crude, outside North
America. (That non-American production in 1972,
incidentally, was not far short of double these
majors' total crude availability in 1982).

Participation from 1973 onwards, which reached
100 percent by 1976-77, in principle and in
financial effect transformed almost all liftings
from most OPEC countries into arm's length
transactions. From 1975 onwards, moreover, the
governments made sure that integrated offtakers
should not get more than small margins of the
economic rents on crude that had been so greatly
increased in 1973. Even so, in practice, most of
these new arm's length sales were still to

integrated offtakers at somewhat preferential
prices, usually with some guarantee of continuing
volumes. Most of the main OPEC exporters sold only
limited volumes directly through their national oil
companies. Also, although governments were setting
maximum production allowables, in some cases that
left the main offtakers' liftings as the
determinants of actual current production.

    The Iranian revolution and its aftermath of
soaring prices effectively did away with most of the
majors' preferential long-term purchases of crude
except through Aramco. Since then, other majors
have had to buy formerly preferential crude only on
the open market terms available to all comers, with
few guarantees.

    By end-1979 only about 15 percent of OPEC
crude exports moved through preferential contracts,
almost entirely to the ex-shareholders of Aramco.
The preference those companies enjoyed had its value
fortuitously inflated in 1979 and in 1980. Then in
1981 and 1982 it lurched into a heavy minus value.
By early last year, most of these companies had cut
their liftings from Aramco down sharply. Their
remaining margins below GSP were not enough to make
the crude commercially attractive. And the Saudi
Arabian government, which had been relatively slow
to push its direct sales, no longer appears content
to rely as much on those offtakers as before.

    During 1979 and 1980, when most of the majors'
preferential deals were being discarded, another
kind of preferential sales from OPEC countries rose
to a peak. Those were government-to-government
deals. They were often not preferential in price,
though at times they were in allocation. Often the
prices involved were not clear, there were elements
of aid, or there looked to be indefinable quids pro
quo involved. To class these as "preferential" is
simply to separate them from sales on open market
terms available to all comers.

    In those years, government-to-government
deals, variously defined, may have reached 20 percent of OPEC's crude exports. With the subsequent
slump in oil demand, they have largely diminished.
But for reasons we can touch on later, they will no
doubt remain a constant element in OPEC government
sales, although fluctuating in volume.

    Today, to sum up, probably 85 percent of the
crude moving in inter-regional trade is sold at
arm's length, and 65 percent of it at prices
available to all comers. Indeed, only about 40
percent of total non-Communist production is equity

crude, and some 28 out of that 40 percent is in North America.

Thus the values at which most crude moves are nowadays prices paid in actual sales. Equity tax-paid costs now apply only to a minority, though an influential minority, of crude supplies. What the industry used to call "realised prices", to distinguish them from the posted prices used for upstream taxation, are now almost all that matters.

Apart from the larger volumes of crude sold at arm's length, the number of transactions for any given volume of crude flows has probably increased too. No national oil company selling its country's crude has the wide menu of different qualities that each of the integrated majors used to be able to offer third party refiners. Moreover, since 1979 term contracts have seldom exceeded a year's duration, and most of those still extant have short phase-out clauses that allow for more renegotiation of price. The number of refiners using crude has not increased. There are fewer - though some subsidiaries of major companies may now have more individual power to bargain. However, there are more professional intermediaries in the trade, and in some regions single cargoes change ownership, on paper, many times.

Hence international crude movement now involves a market of much more continuously active bargaining. That is not to say that this market is perfectly transparent. Term contracts are seldom reported in any detail. Nobody knows with precision the volumes sold under different forms of contract. (Indeed, some of the most important OPEC governments are at pains nowadays to keep even their monthly production figures secret). What has happened is a greater concentration of effort upon reporting spot market prices for crude, and the wider, much faster diffusion of what the price reporting sources are told.

The Significance of the Spot Market

The spot market for crude was not important, or at any rate was not regularly reported, before the late seventies. Such sales served a balancing function at the margin, to match integrated suppliers' crude availabilities with refiners' requirements and meet minor shifts in consumers' demand for products. (In the days of dominant integration, however, exchanges of crude between companies, on a volumetric basis with no prices at all disclosed, were probably more important in matching imbalances in geography or quality).

Discounts reported or rumoured to be available on third party term sales, and the occasional spot sales reported, had some influence even then upon the official prices that OPEC set. But the strongest influence back upstream from any open market probably came from spot products prices and the refinery netback valuations these implied for different crudes. Spot product prices have been reported internationally for much longer than spot crude prices. They have long been accepted as marginal indicators of the current market - for example for escalation clauses in some term contracts, and in certain price control formulae - to a degree that spot crude prices have only very recently.

Changes in those spot products prices in regional refining entrepots represent inescapable signals of strength or weakness in the final market for oil. Any products prices charged in mainstream markets regionally cannot long stay out of line with them. As modified by sizeable excise taxes on some products, these final prices help determine consumers' demand for oil. So they react back on the values that suppliers can put upon their crudes. (This is an interaction, with neither side necessarily determinant).

Spot products prices and refinery netbacks also offer one way, though not the only one, of reckoning crude price differentials. In the trans-formed crude market of today, those differentials have become much more important than they were. When crudes moved mainly through integrated channels, their logistics were programmed according to costs, not prices. Those costs were incremental costs right through the systems - production, transport, refining and distribution, of each integrated supplier. Crude price differentials were not dominant in refiners' or supply planners' choices. Each refiner's own product netbacks and costs, not those of some hypothetical average refiner for a whole market region, determined his choices of crude or contributed to optimising each supply programme. Even the third party refiner was negotiating his own bargain,and seldom paying any published price.

Indeed, then, technical crude evaluations were commissioned as often as not by governments, who were trying to make sure that the tax reference prices reported by their exporting or importing companies were reasonable.

It is still governments who are most
interested in crude price differentials. But no
longer usually for fiscal reasons. The governments
most concerned are those of OPEC. They sell most of
the crude entering world trade today, and they are
mainly non-integrated sellers except in their own
countries. They do not possess the two-way flows of
information, vertical or horizontal, about costs
from well to pump that the formerly dominant
integrated suppliers had. Their profits are made at
the point of crude sale, no later. That would in
any case make them interested in getting crude
prices "right". Moreover, when, as now, they are
seeking to control each other's production in a weak
market, it becomes even more desirable for them to
get the relative prices of all the different crudes
"properly" set. To say that this is "right and
proper", of course, does not necessarily mean that
it is practicable.

NEW MARKET PARTICIPANTS

So here we have a crude market that has been widely
unfolded, and on the buyers' side has become very
competitive and more (though far from fully)
transparent. But the main sellers, as well as
suppliers, are now governments. And as market
participants, the OPEC governments appear to me
inherently different kinds of economic animal from
all other traders. Even in theory I doubt whether
economists would expect their motivations to be the
same as those of private producers for sale, or
other traders. In practice, their opportunities and
the policies they have chosen differ even more.
The economic interests of any government are
national, not multinational. They do not pretend to
optimize at any "higher" level. Each may recognize
increasing interdependence in the international
economy or, for example, common interests within
OPEC. But those are simply external factors it
bears in mind in seeking to optimize its own
national performance, economic and other. These
particular governments, moreover, have generally
monoculture economies. They have little national
income other than their own petroleum revenues and
some petroleum-related industry and commerce. But
they own absolutely the oil that they produce, and
the underlying reserves. They pay no taxes. (Some
of these OPEC governments, indeed, levy hardly any
taxes either, save on petroleum. That increases
their oil dependence). So they do not have to share
their profits and whatever economic rents they can

gain with any other authority.

Theoretically, it is often argued that governments have different rates of social time discount from private entrepreneurs appraising similar projects. Vis-a-vis development of petroleum production, they certainly do have different time horizons from those of private equity producers, whose access to reserves in most countries outside the USA is time-limited by contract. The governments can choose their own depletion rates in developing national reserves.

As for development outside their own frontiers, they have much less incentive to plough their oil profits and rents back into exploration risks than the majors had (and still have, in areas they consider politically safe). Nowhere abroad may look politically safe to an OPEC government for exploration (though some are venturing). These governments today have huge spare capacity. They have more than enough reserves of their own for the moment, and their internal acreage is arguably more "prospective" than anywhere else in the world. Abroad, they would have to explore under other governments' sovereignty. In most circumstances that one can imagine, those other host governments would probably press any explorers from OPEC who discovered petroleum to develop it even if it was high in cost - in order, of course, to save imports from OPEC.

There would be differences in attitude, as suppliers, to investing in further oil production. In a period when little development of production is going on inside OPEC, these differences have hardly shown up. Considering that existing producing capacity in OPEC is surplus not only to current demand but to the production allowables that most OPEC governments have ever set out as desirable, their producing operations need little more than investment merely to offset decline.

Already, these OPEC governments have developed more refining capacity for export from their countries than private operators producing there probably ever would have. Adding the value, employment, and related skills in their economies rather than elsewhere is a legitimate national consideration that might not in principle apply to multinational investment decision-makers. Gas that would otherwise be flared, at cost, is cheap refinery fuel. There may be some considerable economies of scale to be gained through large products tankers; and transport costs, or any kind

65

of costs for that matter, are now far smaller in
relation to the prices being obtained for products
or crude. Politics will certainly have a lot to do
with these decisions about where refineries "ought"
to be put. During the postwar era, they usually
have had.

How far the OPEC governments acquire or
develop refining and marketing downstream abroad is
a crucial question for the whole oil industry in the
eighties. But the answers that emerge need not be
much affected by these producers' being governments,
or in these areas of decision be any different from
private entrepreneurs. As crude-rich exporters with
supply costs far below the going prices of
competitive energy, some of them may be tempted to
acquire refining and marketing facilities abroad
simply to move extra crude that gains them high
profits and economic rents. So were the companies
that developed most of those facilities originally.

The Difference in Marketing Opportunities

Whether these OPEC governments' investment decisions
about production and refining differ radically from
those that private companies in similar (never the
same) circumstances might have made can only be
judged in some years' time. But in marketing now,
their opportunities and hence chosen policies
already differ demonstrably from those of private
companies that might have been exporting the same
crude. Their economic interest in any sale
differs. To begin with, high price levels for
crude, at least in the short run, pay them far more
than private equity producers. In the long and even
medium run, it may be argued, OPEC governments will
face the same effects of demand elasticity to price
as will private producers. But in the meantime,
each dollar by which the crude price rises or falls
means a full dollar of present value to them. To
the equity producer paying around 85% tax, the same
change means only 15 cents up or down.

Even Professor Morris Adelman, who may regard
any such distinctions between governments' and
entrepreneurs' market behaviour as somewhat
irrelevent, does recognize one significant different
of opportunity for these new government sellers to
achieve or sustain higher crude prices - one extra
liberty that governments can take. No anti-trust
laws exist internationally to discourage them from
forming cartels. That indeed may be convenient.
But I myself doubt whether national sovereignty, as
such, has many other particular advantages for

cartel organization. By definition, governments can
hardly be disciplined. In practice, their marketing
behaviour appears to be very difficult even to
monitor.

Theoretically and in practice, OPEC government
sellers do seem more averse to instability in the
crude price than private traders. (Both, on
balance, have not done too badly out of price
instability during the last decade). Certainly an
almost complete dependence upon oil revenues - the
monoculture vulnerability again - makes these
governments' planning hazardous when prices look
unstable. And unless they can achieve other
development plans, they may never escape that
dependence. So far, this OPEC aversion has been
manifested mostly towards instability downwards.
But I don't think there is too much ground for
cynicism here. Most OPEC governments seem genuinely
reluctant to see any sudden upsurge in prices again,
for example through some Gulf political or military
crisis. Such crises would affect some of them
directly, but they are worried by the medium-term
economic implications too. The price ratchet might
not apply next time. After any emergency was
remedied, the subsequent price slide might steepen
towards a collapse (which is highly possible
anyway).

This desire for stability, and the yearning
for "price administration" by cartel organization or
other means, tends to align OPEC interests as
sellers against the new forces from the buyers' and
traders' side that are now tending to open up the
crude market. Yet it was primarily OPEC action in
1979, cancelling long-term preferential contracts
and fragmenting the crude trade, that widened the
scope of the spot physicals market in crude.
(Rather ironically, today, about the only fairly
reliable evidence that OPEC's monitors have of
whether their members are in fact holding their
prices or not is whether spot market prices for
crude diverge significantly from official levels).

That cancellation of long-term contracts
turned the ex-integrated majors into net buyers of
crude. Though they remain substantial equity
producers with some share of economic rents on crude
to consider, some of them are making profits downstream and all have to hope they eventually will.
Their concern not to see crude prices slip is no
longer as unmixed as that of OPEC (and some other
exporting) governments. But, for the moment, shared
fear of that collapse unites their inhibitions.

Fragmentation of the crude market in 1979-80 attracted in many additional traders from other commodity markets. The subsequent weakening of demand, which made crude purchasers in their turn uninterested in term contracts at official prices, has tended to link more and more of the deals that are made with spot prices, if only in shortened phase-out options. The speculative gains that some traders made in 1979-80 recreated interest in futures markets for gas oil and more recently for crude. Futures markets are of course the antithesis of what many established operators in this trade, OPEC or private, consider respectable. Beyond their "acceptable" functions of hedging, they draw upon speculators concerned simply with the course of price expectations, not with the physical commodity at all. Both functions, in theory, should help to strengthen what might be called the "price auction competitiveness" of the oil market. They may also have rendered it more transparent. At any rate, futures markets seem to act as a kind of lens, focusing instant attention upon whatever occurs, or is rumoured to have occurred, that might affect spot markets for oil. In spite of the apparent failure of the London crude oil futures contract, a good deal of forward trading in "paper barrels" is nevertheless taking place in the spot market for North Sea crude, and actual traders in "wet barrels" have to take notice of this phenomenon whether they like it or not. Eventually, whatever their inhibitions, OPEC governments may find themselves obliged to play a role as well.

It is still very difficult to judge whether one is getting more information nowadays about the physical spot market for crude, or simply getting the same limited information faster, spread more thinly across more services flashing onto one's computer monitors. (The futures markets, of course, are ideally transparent. Every bargain is recorded instantly and irrevocably. The only thing one does not know is how many of them are directly related to the oil business).

Both these open markets, also, have been criticized as reflecting too strongly what happens in two regions that are particularly prone to short-term dealing, the North Sea and the US Gulf. In terms of the concentration of price reporting, that may be true. But in terms of reflecting major influences upon world oil prices everywhere, I doubt whether this focus of attention is exaggerated. Those two regional markets for non-OPEC crude, in

these current years, may well mount the strongest
demand-side influences upon the world crude market.
The pressures on price with which OPEC production
allocation will have to cope - and so far,
admittedly, has coped - first come to the surface in
those non-OPEC crude markets.

So, we have looked at two aspects of the same
sea-change in this international industry. One is a
change in the matrix of transactions across the
trade. The other is a new set of principal actors,
in my view distinctly different from all other kinds
of actor. The influences of these two changes are
inter-acting and to a considerable extent
conflicting. But they are not the basic
determinants of market performance. Those are still
underlying demand response to price - which in my
view no energy economist yet understands with any
precision - general economic performances in the
Western and newly industrialising economies, and
other external influences - like the weather.

5 OPEC AND STRUCTURAL CHANGE

Ian Seymour

Given the sheer magnitude of the changes that have
taken place in the oil industry in recent years –
which have been outlined in detail in the papers by
Paul Stevens and Jack Hartshorn – it is hardly
surprising that OPEC as the residual supplier of
energy to the world should have found itself at the
sharp end of these profound structural shifts. On
the one hand, following the price shock of 1979/81,
there has been the devasting deflation in demand for
OPEC oil from 31 million barrels a day in 1977/79 to
17.5 million barrels a day in 1983, combined with a
commensurate erosion of OPEC's percentage shares of
the oil and energy markets: as a proportion of
non-Communist world oil production, down from 63 per
cent in the early 1970s to 42 per cent in 1983, and
as a proportion of non-Communist world energy
supplies, down from 39 per cent in 1973 to 19 per
cent in 1983.

Another galvanizing force for structural
change has been the fragmentation of the world oil
industry following the assumption of control over
decision-making on crude oil prices and production
volumes by the governments of the oil exporting
countries, thereby precipitating the de-integration
of the monoliths which had so dominated the oil
industry for the previous half century.

These two momentous trends have forced the
OPEC countries to undertake a radical rethinking of
all the traditional assumptions that have guided
their past policies and actions. This jettisoning
of traditional attitudes has been evident both on
the macro level of OPEC thinking on prices and
production, and the interaction between the two, and
on the micro level of the oil strategies of
individual OPEC governments.

70

OPEC AND PRICE/PRODUCTION STRATEGY

From its inception in 1960, OPEC's raison d'etre was based on the objective of defending and improving oil prices. Before 1982, the proposition that it might, in certain circumstances, be in the oil exporters interests to lower prices as well as to raise them, would have been regarded as something akin to treason in OPEC circles. Up to that time, disagreements on prices within OPEC, and there were many, centred on the extent and timing of price increases, not on the principle. Thus, OPEC's first blueprint for long-term strategy which emerged in 1980 after several years of debate, while recognising that everything possible should be done to avoid sudden jumps in price, advocated full compensation for inflation and a steady increase in real prices in line with the economic growth rates in the main industrialised countries.

Saudi Arabia had always been fearful of the repercussions on demand of the runaway price explosion of 1979/81, but it was not until 1982, when the hangover effects of the price spree became painfully apparent, that a significant number of other OPEC exporters began to share Saudi Arabia's apprehensions in a meaningful way. Naturally enough, such fears were more in evidence among the producers of the Arabian peninsula whose enormous oil reserves made them more anxious to preserve oil's position in the overall world energy balance than to maximise immediate financial gains. And so the scene was set for the deliberate and orderly, or at least fairly orderly, $5 a barrel reduction in OPEC's official selling prices which was enshrined in the London agreement of March 1983. This was partly in response to what had already happened in the market, and partly a measured recognition that the previous price rises had gone too far for OPEC's good.

Owing to a consensus among the main exporters – and despite opposition on a theoretical plane from the remaining price hawks such as Iran – a new de facto OPEC strategy has emerged. This is based on the premise that, although price moderation is urgently needed to restimulate demand, no further reduction in nominal prices can be contemplated since this would be too destablising for the financially disadvantaged OPEC countries. Therefore, it is held that there should be a lengthy freeze of prices in nominal terms, allowing for erosion in real terms, until such time as demand recovers sufficiently to reach what is described as

a level of comfort for OPEC members, estimated by
some at around 24 million barrels a day. After
that, perhaps in 1987 or so, some price increases in
nominal and perhaps real terms might be
contemplated. Now, as yet, this strategy has not
been given a chance to work, largely because of the
strength of the US dollar on the financial markets.
This has had a positive side and a negative side as
far as OPEC is concerned. On the positive side, it
has cushioned the adverse financial effects of the
price reduction for those OPEC countries which are
in financial need and made life considerably easier
for them. On the other hand, it has largely negated
OPEC's strategy in so far as the erosion of prices
in real terms is concerned since, in a great number
of markets, Europe in particular, the reduction in
dollar terms has been almost entirely negated by the
rise in the value of the dollar in local currency
terms. The crisis has also had the effect of
resuscitating after a long period of, what was one
of OPEC's earliest ideas, namely the pro-rating of
producton among member countries for the purposes of
market stabilisation. As you will recall, the
founding fathers of OPEC and the redoubtable Paris
Alfonso in particular, always advocated this, but
the idea lapsed owing to opposition from the Gulf
producers who did not want to see any interference
with the development of nascent oil capacity. So,
the question was resurrected at various times during
OPEC's history, but it always came up against an
absolute veto from Saudi Arabia. However, the
danger posed by the market collapse in 1982 and late
1983 was such as to convince even Saudi Arabia of
the necessity of using production programming to
underpin any market stabilisation plan.

Nevertheless, in my opinion at least, it is
not the actual production programme which has been
responsible for the relative success of OPEC's
market stabilisation efforts since March 1983.
Basically, I think it has been much more a matter of
price discipline. That is to say that the OPEC
countries have realised that it is simply not worth
the risk of massive discounting in order to gain
significant increases of volume at the expense of
other producers, which some of them were definitely
doing in 1982 and early 1983. Since March 1983, the
price discipline has been much better. There have
been violations, but these have been on a relatively
minor scale and I do not think one could say that
any OPEC country has been really pushing volume at
the expense of price. So long as price discipline

72

is maintained - that is to say nobody sells oil
below the official price - this provides a sort of
self-regulating mechanism. In such a case, it does
not really matter if quotas are exceeded by a small
amount at any particular time, because the sellers'
refusal to discount their prices will of itself
bring about automatic readjustment of production.

OPEC AND NON-OPEC

Basically there are four factors responsible for the
spectacular fall in demand for OPEC oil in the
aftermath of the price shocks: conservation in
energy use, the effects of the economic recession,
substitution of oil by alternative fuels, and the
growth of non-OPEC oil supplies. OPEC price
moderation could have some influence on the first
three factors but is unlikely to have any impact on
the fourth, the presumption being that by and large
non-OPEC oil will always be produced at the
maximum. So far, the non-OPEC exporters have been
able to have their cake and eat it too, sheltering
under the umbrella of the OPEC price structure and
volume restraint while pushing their own production
to the maximum.

In this connection it was somewhat ironic to
hear Mr Buchanan-Smith complaining last night about
"unfair" competition from future exports of refined
products by OPEC countries. At the same time, he
omitted to mention that the UK has been maximising
its crude oil output at the expense of OPEC for some
time past. Whereas OPEC production is now operating
at little over 50 per cent of capacity, there is not
a single barrel of spare capacity in the North Sea.
In practical terms, probably there is not very much
that OPEC can do about this, short of allowing a
price war to develop to the detriment of everyone.
Realistically, probably, the main hope of the OPEC
countries is that output from the main centres of
non-OPEC oil whose reserves are not so prolific will
shortly peak and begin to decline.

OPEC AND THE DOWNSTREAM

As one would expect, the profound changes in recent
years have also had an impact on OPEC's thinking in
general, and that of individual OPEC countries in
particular, about the downstream. This is not so
much a question of any response to structural
change, but simply a natural variation of response
in the face of firm or soft market conditions. When
market conditions were soft in the 1960s, OPEC
countries were preoccupied with security of outlets

for their crude and thought of going downstream
outside their borders as a means of achieving this.
At that time, in the mid-1960s, Shaikh Yamani
espoused the idea of a close linkage between Saudi
Arabia and the major US oil companies in Aramco,
which would start with Saudi government
participation in the upstream operations and then
gradually spread to the downstream, so that the
government would be a partner of the majors both
upstream and downstream in the market. However,
although a degree of government-company
participation was instituted in the upstream in the
early 1970s, it was fairly rapidly overtaken by
events in most cases; and in the downstream it never
got off the ground at all.

In the early 1970s when the oil market firmed
up considerably and the questions became not whether
oil would find markets but whether markets would
find oil, OPEC thinking on downstream involvement
underwent a considerable change. On the one hand,
because of the strength of the market, the OPEC
exporters saw little reason to invest in downstream
operations abroad - they felt no problem about
finding a home for their oil. On the other hand,
they did want to expand their share of world
refining because they felt that there was value
added to be had there, if not in actual cash terms
then in terms of industrialisation, employment
opportunities and technological advancement. Also,
some countries - Kuwait is an example - wanted more
domestic refining capability in order to upgrade what
was essentially a rather poor quality of crude and
thereby enhance the market prospects for that
crude. The upshot of all this was that most if not
all OPEC members decided to confine their refinery
expansion activities to their own countries and a
number of them - Saudi Arabia, Kuwait, Algeria and
others - set in motion very ambitious plans for the
construction of new refining capacity for export.
And it is these new refineries that are now coming
on stream, much to the consternation of Mr Buchanan-
Smith and others.

But in this sphere, too, times have changed.
In the soft market that we have today, the bulk
marketing of refined products for export, from
export refineries in the producing countries, does
not in itself confer any more security of outlet
than the bulk marketing of crude oil. To gain this
type of security of outlet, you have to go further
downstream to distribution networks in the main
markets overseas - service stations, dealerships

etc. There has lately been a definite renewal of
interest by some OPEC countries in this type of down
stream penetration into consumer markets. But so
far only Kuwait has made any practical and
successful headway in acquiring and operating
refining and distribution facilities overseas.
Personally, I feel that the scope for such
acquisitions by OPEC governments in the future will
be rather limited. Successful downstream
acquisitions require the presence of a rather
special set of conditions in the OPEC country
concerned. Firstly of course, it must have the
requisite spare cash for such acquisitions.
Secondly, it must have a complete and unconditional
political endorsement for such a policy at the top
level. Thirdly, there must be the ability of the
top executives of the OPEC national company
concerned to negotiate and finalise international
takeover deals with the requisite degree of
confidentiality, expertise and sheer commercial
flair. And fourthly, the national company concerned
must have the capacity to absorb, manage and direct
the new systems once they are acquired. Now, these
conditions happen to be fulfilled in the case of
Kuwait, but elsewhere circumstances are much more
problematic.

Of course, the really important question mark
hovers over Saudi Arabia. Will Saudi Arabia go the
same way as Kuwait in acquiring distribution
networks downstream? This would clearly have very
profound consequences for the oil market.
Obviously, one can see circumstances where such
acquisitions might perhaps be very much to Saudi
Arabia's advantage. But we must remember that
changes of direction in policy matters often move
rather slowly in Saudi Arabia. In the mid-1970s a
firm policy was taken in Saudi Arabia to locate all
future refining activities in Saudi Arabia itself.
Now, in Saudi Arabia, as elsewhere, it takes quite a
long time for a policy to be finalised and accepted
and it also takes quite a long time for a policy to
be changed. However, there are signs of change in
Saudi Arabia. Recently, Shaikh Yamani disclosed
publicly that King Fahd had in fact endorsed in
principle the idea the Kingdom should investigate
the possibility of acquiring downstream facilities
overseas. Shaikh Yamani said that he himself was in
fact in favour of such a policy and he commended
Kuwait's existing acquisitions. But he did at the
same time inject a note of warning into this, saying
that in fact a lot of refineries and facilities on

offer were nothing more than obsolete scrap metal as
he called it.

OPEC AND CHANGES IN MARKETING PATTERNS

I would like to dwell briefly on the effects that
the structural changes have had on marketing
patterns in OPEC countries. Jack Hartshorn has
already described how the old long term contracts
were broken down under the impact firstly of the
takeover of decision-making by the OPEC governments
and then under the impact of the Iranian crisis of
1979/81 (see pages 57 - 67). At that time it was
certainly the OPEC governments themselves that dealt
the death blow to what remained of the old long-term
contract system. In the heat of the crisis, under
government fiat, prices were changed, volumes were
cut back and so forth. And at the end of it all,
perhaps not surprisingly, buyers felt that there was
nothing to be gained in maintaining a long term
contract in the future. It made you pay more when
the market was weak and did not protect you when the
market was strong. Now, typically, that old rigid
type of long-term contract has been replaced by a
term contract which is usually a sort of 12-month
evergreen type of deal with appropriate price
reopeners and phase-out provisions.

Beyond these 12-month evergreen term contracts
there is a lot of what has been described as
"imaginative marketing" in the OPEC countries, some
of which is legal under OPEC rules and some of which
is illegal. For example, processing deals based on
the disposal of refined products at realised prices
are clearly illegal within the OPEC rules. However,
the use of crude oil for repayment of commercial
debts formally invoiced at official prices but which
often later surfaces on the spot market at below
official prices, is probably a border line case.
But what have emerged strongly in recent times have
been various types of what might be called "package
marketing", which do not in fact violate the OPEC
price rules in any formal sense. Sometimes these
are crude sales at official prices linked in some
way with sales of condensate or refined products
(whose prices are not bound by OPEC rules) at open
market prices. Then there is the sales package in
which the proportion of various grades of crude at
official prices are mixed in order to offset the
spot penalty against one grade of crude by a premium
on another. In other words, the mixture becomes a
saleable package when the weighted average of spot
prices for the crudes concerned equals or exceeds

the weighted average of the official prices. This
method is one that has recently been used by Saudi
Arabia's Norbec for third-party sales, but there are
examples of the same sort of packaging in other OPEC
countries.

Partly, this package marketing is a response
to the fact that differentials between the official
prices of various qualities of crude oil are now
more than ever immovably embedded in the London
agreement and can no longer, even if they ever
could, be freely adjusted from time to time to
reflect changes in relative values. As a result,
differentials have got way out of line, with Gulf
light crude labouring under spot penalties of 50
cents to a dollar a barrel and heavy crudes
benefitting from a corresponding premium.
Consequently, producers of the heavier grades have
few marketing problems and those who produce a
variety of grades can balance one off against the
other. But those in the Gulf who produce only light
crude are having a tough time of disposing of their
oil at official prices. This is a problem that will
not be resolved till OPEC can once again get to
grips with the perennial bugbear of differentials.

OPEC AND THE TRANSPORT SECTOR
In addition to the changes in the marketing sphere,
I would like to mention the effects of recent shifts
in another area, namely transport. It was long an
aspiration of the OPEC countries to build up their
own tanker fleets as a means of enhancing their
security of outlet and playing a larger part in the
oil industry in general. Now, of course, the
collapse of demand in the past few years has had an
absolutely devastating effect on the plans of the
OPEC countries to build up their crude oil tanker
fleets. In some cases, like LPG and product
carriers, the build-up has continued, but as far as
crude oil carriers are concerned, it has been a
disaster. For example, many of the tankers owned by
the OAPEC-sponsored Arab Maritime Petroleum
Transport Company (AMPTC), which are mainly
newly-built vessels, are now being sold for scrap
although some of them have have never been used.
But even this cloud has some kind of a silver
lining, and in some cases OPEC countries have
introduced fruitful transport innovations. For
example, Kuwait's KPC, which has a rather individual
mix of crude and products in its transport pattern,
has pioneered the use of VLCC tankers with
segregated tanks for carrying cargoes composed

partly of crude oil and partly of products, with considerable overall cost savings. So the transport outlook for the OPEC countries is not altogether bleak because there are areas in which special systems and special requirements can make use of the opportunities afforded by the present situation, though the previous idea of a huge build up of Arab or OPEC-owned tanker capacity is obviously no longer feasible.

MORE REALISM IN OPEC
In conclusion, I would just like to say that these structural changes, painful though they may be, have not been without their beneficial side as far as OPEC is concerned. There is now more realism in the face of international oil problems and more flexibility of response. Also, the OPEC countries are learning to manage their economies better; they are learning to make do with less money. They are beginning to rationalise their economies on a sane basis which was not happening during the period of the massive increase of revenue. So perhaps this cold shower of the oil glut and falling prices has not been without an overall benefit for the economies of the OPEC countries. I personally believe that there is no problem of an economic nature that the OPEC countries cannot handle. The problem for OPEC is really political: the fact that there is a major war between two of the main member countries and the fact that the Middle East is an area plunged into potential chaos and instability, not only the Iraq-Iran war but by the Israel-Arab conflict and the chaos in Lebanon. These are problems which really have tended to paralyse any decision about future strategy in OPEC. Perhaps the miracle is that they have been able to do as much as they have, that Iran and Iraq, two countries at war, have been able to sit side by side in OPEC conferences and actually reach agreement on essential matters of price when their vital interests on oil prices were seriously threatened. It just shows that there is a bedrock of cohesion in OPEC which is sufficient to defend the basic level of oil prices. But when it coms to anything more than that, when it comes to any look at future strategy, the decision making in the organisation tends to be paralysed because there simply is not the rapport any longer between the member countries. And this is a political problem.

I do not intend to go any further in the political field and examine the possible

implications of any significant disruption of Gulf
oil supplies though the Strait of Hormuz. But I
would just like to retell an anecdote about a recent
conversation of mine with a top oil official in one
of the leading third-world oil consuming countries.
He was talking about his fears about an explosion in
the Gulf and a third oil price shock as a result and
he said: "Well, I realise there may well be a third
oil price shock but I assure you that if there is,
that will be the last, because if there is another
price explosion, the interests of the consuming
governments, both in the industrialised world and
the Third World, in relegating OPEC oil to a minor
role in the world energy pattern will become so
overwhelming that no amount of assurances of price
moderation on the part of OPEC will be enough to
dispel it". This, I think, should give everyone of
us and in OPEC food for thought.

6 DOWNSTREAM IMPLICATIONS OF STRUCTURAL CHANGE

Edith Penrose

In this paper I shall be primarily concerned with the implications for the downstream of the international oil industry of the "structural changes" that have taken place, or still are taking place, upstream. But before the subject can be seriously analysed it is necessary to ask what is meant by "structure" and "structural change". How should we distinguish structural change from other kinds of change?

The word "structure" has many meanings, but in general it carries the connotation of a framework within which activity is carried on and which is itself not easily changed. It is a framework relatively invariant over time; substantial changes do not take place quickly as a function of normal on-going activity,[1] although over a "long" period such activity may bring about a structural change. So long as change is marginal, slow, continuous, reasonably predictable from period to period and does not create difficult adjustment problems for continuing operation, we do not notice that a structural change has taken place until after the form of the new structure becomes visible. In historical perspective almost nothing can be looked on as "invariant"; all structures change over time. Rapid and discontinuous change is another matter. It is this kind of change with which we are concerned here.

Many, but not all, of the structural changes that have occurred upstream in the past ten years would probably have occurred over 15 or 20 years in any case, but had they not taken place in the context of wars, civil disturbances, severe economic recession and inflation, and embedded in the international tensions of the Middle East, their impact on, and implications for, the downstream

80

would almost certainly have been very different.
This implies that not only was structural change
itself relevant, but also the conditions under which
it occurred. The downstream has been reacting to an
international oil industry still in a state of
precarious disequilibrium; it is not unlikely that
the upstream-induced downstream changes will in turn
bring about further structural change upstream. It
is not at all clear what kind of structure the
industry will settle down with.

We are thus presented with a major difficulty,
for in order to analyse the implications or effects,
of a given cause, in this case a structural change,
we have to assume that the reference structure is
given for the period to which the analysis is
intended to apply. We can discuss the observed
effect of what has happened downstream and try to
discern what is likely to remain relatively
unchanged for the period we want to consider, but it
is, in a sense, too early to deal seriously with the
implications of structural change for any but a very
short period say, the next five years, or up to
about 1990.

The major change upstream on which most
attention has been concentrated is the change in
what might be called the administrative structure of
the industry which was brought about by the
assumption of ownership and control of upstream
operations by the major oil-producing countries of
the Middle East - the so-called "dis-integration" of
the industry. This "dis-integration" however, is
already being redressed in some degree as downstream
activities move to oil-producing areas where the
governments continue investment in refining and
petrochemicals and begin to develop export markets
for their new products. There is also some movement
from oil-exporting countries into refining outside
their borders. Nevertheless, it will be a long
time, if ever, before this "re-integration" will
seriously begin to displace the new international
market for crude oil.

It has been widely held that extensive
integration, particularly integration between crude-
oil production, refining and distribution, is a
necessary condition for the efficient operation of
the industry and for the maintenance of reasonable
stability. Integration may provide assured outlets
for crude oil, leading to steadier and more
efficient planning of output over time, may permit
more efficient operation of refineries as a result
of an assured and managed flow of crude and a more

81

flexible adjustment to short-run changes in demand for different products in different areas, thus avoiding disruptive fluctuations in prices which would raise costs to both producers and consumers.

Historically, much of the reason for the vertical integration of the industry related to the monopolistic nature of competition: refiners often sought own sources of crude oil because the market for crude was imperfect and dominated by a few companies; crude-oil producers often sought owned outlets for their crude to protect themselves against their rivals and to control markets more effectively. Whether or not vertical integration was a particularly efficient way of organizing the industry, it became a widespread competitive necessity for at least the leading firms, its existence providing its justification.

It has also been argued that vertical integration is necessary to stabilize the industry in view of the uncertain results of exploration, which may cause prices to fall heavily in the face of spectacular success and to soar when exploration fails to maintain adequate P/R ratios, the leads and lags producing alternating glut and scarcity. This argument was originally put forward by Paul Frankel and subsequently countered by Morris Adelman, who stressed the role of the cost of developing discovered reserves and insisted that even in competitive markets companies would not develop reserves without reference to the effect on profitability and at prices unrelated to replacement costs.

It can hardly be denied that the disintegration of the industry has been accompanied by increased instability, as Frankel would presumably have predicted. Nor can it be denied that it has not been replaced by free competitive markets for crude oil, which provided the framework for Adelman's analysis. Paul Stevens in his paper to this conference stressed the fact that the vertical link between the downstream and crude-oil production was instrumental in containing the pressures on oil markets of the extensive excess productive capacity and thus helped to maintain product prices and the value of crude oil. Similar efforts to contain excess capacity are now made through a different mechanism, of which OPEC is the centre. Stevens also noted, as did Jack Hartshorn in his paper, that the efforts of the OPEC governments to control oil output are handicapped by an inadequate knowledge of what its numbers are in fact producing. When the companies were in place, the considerable knowledge that each had of the production and plans of others helped to restrain oligopolistic competitive behaviour, at least within limits. This was largely a result of their extensive horizontal geographical integration in the production of crude oil through their jointly owned consortia.

The break-up of the vertical and horizontal integration of the companies naturally led to a change in market structure downstream as well as upstream since by its very nature the market lies at the interface of the two levels of the industry. A rise in arm's length dealing was to be expected. Most of the transactions between the upstream and the downstream of the industry that had previously been internal transactions within the integrated companies have now become externalized, and Mr Reid gave us in his paper a lively picture for one large company of its new relationship with the market.

Other implications of the fact that the companies are now net buyers of crude, that the role of the spot market has become dominant, that traders, speculators and commodity brokers are of increasing importance, that an emerging futures market might help solve some of the problems associated with the pricing of crude oil, have all been widely discussed. There is little I can add here. At the same time, however, there seems no convincing reason why the change in the administrative structure of the industry should by itself have brought about the virtual replacement of long-term contracts by very short-term ones; no inherent necessity for the dominance of trading on the spot market. In different circumstances the oil companies might have replaced their equity liftings with liftings under lòng-term contracts and negociated arrangements.

It is, therefore, perhaps interesting to ask how far the changes in downstream markets and activities are really attributable exclusively to the reduction of integration with the upstream. After all, a much wider and more complex set of changes have occurred in the relation between the upstream and the downstream of this industry. Some of these changes were associated with the inter-action between general economic developments as they affected oil markets, and the specific economic and political relationships among the OPEC countries themselves. Some of the latter were simply the side-effects of political events unconnected with

the oil industry. It would, I think, be misleading
to attribute all of the observed changes in
behaviour and structure downstream to the change in
the upstream position of the oil companies,
although, to be sure most of the market changes
would not have occurred in the absence of the
breakdown of company integration. In other words,
this change in structure was a necessary condition
for the emergence of the market as it in fact
appeared; it was by no means a sufficient one. The
nature of activity and the relationships in crude-
oil markets as they in practice developed were
shaped by other considerations than the mere change
in administrative structure.

An important part of the framework, broadly
interpreted, within which an industry operates is
defined by the conditions of supply and demand as
perceived by buyers and sellers to the extent that
these are expected to persist over the medium term.
The special characteristics surrounding the process
of administrative change were such as to bring in
its wake further upstream structural changes
affecting the supply of crude oil, which in turn
brought about structural changes in demand.

First, the abrupt rise in prices following the
Arab-Israeli war increased the expected profitability of oil reserves in the North Sea, Mexico and
in some other developing countries, as well as of
the development of non-conventional sources of oil
and other forms of energy. Although crude prices in
real terms subsequently declined, causing many of
the new projects to be abandoned, the major non-OPEC
sources of oil and some of the other sources of
energy continued to be attractive. A "structural
change" in the distribution of oil supplies was
affected, which would probably have occurred in any
case but very likely at a slower rate. For much the
same reason technology, especially for off-shore
activities, developed rapidly. This too can be
regarded as a structural change. The implication
for the downstream was, of course, the obverse:
increased prices not only affected the amount of oil
demanded but shifted the demand curve for oil by
inducing consumers to "conserve" oil, not only
through the substitution of other sources of energy
but also through investment in more energy-efficient
buildings and industrial machinery, much of which is
probably irreversible and thus implies not only a
movement down the demand curve but a shift of the
curve itself.

For some years before the first "oil crisis"

there had been signs that most of these changes were already beginning to emerge. The desire on the part of the oil-exporting countries to assure their sovereignty over their own oil industry had long been expressed and there was every reason to expect that in time a change in ownership and control of oil production would be successfully accomplished. Similarly, the development of the North Sea and further development of Mexican oil production, although probably accelerated by the rise in prices were already on the horizon. These sources of non-OPEC oil would have had a similar, though much milder, effect on the demand for OPEC oil if the widely expected scenario of continually rising demand pressing on diminishing oil reserves and leading to price increases during the decade of the 1980s had taken place, instead of war and sharp discontinuity.

Similarly, given the same scenario, demand should have been somewhat checked by the price rises. We can add to this, the effect of the prolonged periods of recession that have characterised the world economic scene over the past decade, recessions which may have been aggrevated by the drastic change in the energy scene but which almost certainly would have occurred in any case and led to a decline in oil consumption.

The increase in prices was, however, not the only consideration. Perhaps of equal importance in assessing implications for the downstream was the sharp rise of uncertainty and the intensification of feelings of insecurity on the part of both industry and governments. Investment in other sources of energy and in oil from other sources than OPEC bought a greater degree of security. The increased uneasiness of consumers about relying on OPEC oil arose not only because they feared that political and military disturbances might disrupt the physical flows of oil, but also because they were uncertain about the intentions, policies and actions of the oil-exporting countries with respect to supply and price. Jack Hartshorn, discussing the effect of taxation on the profitability of equity crude, noted particularly the fact that the disintegration of the industry was not simply the dispossession of the oil companies, but a shift of control from private to public hands. This observation is of great importance, for both the objectives of governments and the processes of decision-making are often very different from those of private industry. This is especially true of a large international industry of

interest to numerous governments and composed of
very large firms.

In almost all countries governments tend to
intervene in "commercial" decisions when they think
it desirable in the "national interest" - a very
elastic and imprecise term. With respect to an
industry as important as is the oil industry for the
major exporting governments, there can be little
question that political considerations are likely to
play an enhanced role in decisions on price and
supply. The unease that this may induce in
consumers is aggrevated when such decisions are
made, not just by individual governments, each
deciding independently where its own interests lie,
but by governments acting more or less in concert,
each forced to make compromises with the others and
arriving at collective decisions through a bargaining process. This increases uncertainty, not only
about the political policies of the producing
governments, but, and perhaps even more acutely,
about their collective capacity to implement their
decisions.

In the face of excess capacity upstream and
sluggish demad downstream the price of crude oil is
at present precariously maintained through the
observance of "price discipline" by sellers of OPEC
and non-OPEC crudes alike. All have an interest in
preventing a "collapse" of oil prices and all fear
such an event if cracks appear in the retaining
wall. The discipline is maintained through a
combination of the explicit regulation of output by
a cartel of oil-producing countries organized in
OPEC and cooperative restraint of major non-OPEC
producers, mainly in the North Sea and Mexico and
supported by large buyers. This loose arrangement
has worked reasonably well up to the time of writing
in spite of the fact that, apart from the importance
to all of price maintenance, the national interest
of each of the countries differs considerably,
particularly with respect to the level and structure
of prices and the allocation of permissible output.

When the international oil companies were in
position upstream they, too, had to take account in
their decisions on price and output of host governments' expression of their national interests,
particularly with respect to the allocation of
offtake among the different sources of supply for
each company, but also in the negotiation of tax
prices. Governments had a direct influence in a
variety of other ways on the activities of the
companies. But when the governments themselves took

over control of the upstream, the influence of the
national interest of the producing governments on
their decisions regarding supply and price was
brought to bear in a very different way.

It sometimes seems to be assumed by observers
of the international industry that the oil companies
("seven sisters"), taken together, could in the past
be treated as a coherent single-minded cartel
maximizing their joint profits, and that producing
countries in OPEC can now be treated in the same
way. This is, of course, far from the truth. The
companies certainly had many close connections, both
horizontal and vertical, which gave them a great
deal of knowledge about each others activities and
facilitated a degree of informal (if not formal)
coordination of their respective upstream policies,
but they were not a cartel attempting to maximize
their joint profits, even upstream, and certainly
not downstream. They were, as Jack Hartshorn has
noted, the "business in between", attempting to
balance an innumerable variety of interests in
pursuit of their own objective of making money from
running this international industry. They were
neither perfect monopolists nor perfect
competitors. The individual interests of each
company were different in most respects from those
of every other, yet they could not afford
unrestrained competition.

OPEC is now, on the other hand, a formal
cartel, and within it the individual interests of
each country must now be reconciled to achieve some
sort of working balance. But both the process of
conciliation and the factors affecting the attitudes
of the different countries stand in sharp contrast
to the loose oligopolistic behaviour of the oil
companies during their moment in the Middle East.

The oil-exporting countries have three sorts
of interests in their oil industry. First, as an
industry existing on the exploitation of an
exhaustible natural resource, the governments have
for a very long time seen the assertion of sovereign
control over it as virtually the symbol and touch-
stone of political sovereignty and independence.
The strength of this feeling varied from country to
country, and in some it was of less concern to the
governments than to important sections of their
populations whose political pressure could not be
ignored.

Secondly, as a major, and for some producers
almost the only, export industry in the country, all
governments have for long felt the necessity of

87

ensuring that it was exploited in such a way as to
maximize the revenues accruing to the country.
Thirdly, they have long wanted to integrate the
crude-oil producing industry with the rest of the
economy through the development of industries
supplying inputs to the oil industry or of
industries using oil itself as an output, notably
refining and petrochemicals. In all three respects,
the governments steadily exerted a variety of
pressures on the companies to achieve their ends, at
the end of 1968 announcing their collective
intention of taking over in the near future.

OPEC (and OAPEC) are organisations which, in
their different fields, were designed to assist the
governments individually and collectively in the
achievement of their objectives. Sovereign control
is now no longer an issue, but the maximization of
revenues and the utilization of oil as the basis for
domestically-based industries (including export
refining) remain of great importance to all
countries.

On the other hand, the well known differences
in their economic, political, and demographic
structures are a source of division and strain. The
"low absorbers" who, because of relatively small
populations and limited other resources, are without
the capacity to use productively the large revenues
accruing to them stand in sharp contrast to the
"high absorbers", who feel continually constrained
by inadequate revenues and whose development plans
suffer severely when revenues fall short of expectations. The size distribution of oil reserves
among the countries is in an almost inverse relation
to the distribution of population and other
resources. This situation naturally affects their
attitudes towards the level and structure of prices
as well as towards the production quotas allocated
to each, on which OPEC must agree if it is to act as
a successful price and output cartel. But, of even
greater importance at times, it also affects
attitudes towards the "justice" of the demands each
puts forward in bargaining with OPEC colleagues.

Is it not "fair" that those countries with
large populations and large development "needs"
should receive "adequate" quotas, and/or prices as
high as possible, to ensure the highest possible
revenues? On the other hand, is it not reasonable
for countries with large reserves and a long
production horizon to insist that prices should be
at a level to ensure the maintenance of the long-run
demand for OPEC oil? Are the latter not "right" to

insist that excessively high prices might bring
about long run changes adversely affecting their
industry by reducing the demand for their oil
through substitution and conservation on the part of
consumers and the development of alternative sources
of oil and other sources of energy? The conflict is
fundamental and at times bitter and acute. And if
to this we add ideological and political differences
and the differing perceptions of the legitimacy of
different political regimes, the brew can become
explosive.

At present the expression of differences is
muted, or at least restrained, on the one hand by
the common fear that, if they do not hang together
they will hang separately on a collapsing price
scaffold, and, on the other, by the hope that not
only will the demand for oil rise strongly with
strong international economic recovery but that the
demand for OPEC oil will also grow as production
outside OPEC reaches its plateau, peaks and begins
its decline.

This is a very fragile situation. It is true
that most analysts, though by no means all, are now
expecting substantial economic recovery, but at the
same time they consistently underline the
uncertainty surrouding their expectations; it is
also true that the demand for oil is expected to
respond to rising prosperity, even if to a lesser
degree than in the past; moreover, it is widely held
that by the late 1980s OPEC production from the Gulf
will rise as a proportion of world oil supplies.

Nevertheless, the consumption of oil per
increment of GNP continues to fall and estimates of
oil reserves outside the OPEC area continue to be
revised upwards. Few have a great deal of
confidence in the optimistic scenario for OPEC, in
the medium term in any event, and even fewer are
prepared to put reasonably firm dates on the points
where a clear change in the present downward
pressures will lead to a sustained upward movement.
Peter Odell and Kenneth Rosing of the Centre for
International Energy Studies, Erasmus University,
Rotterdam, have produced estimates of the long-run
supply curves of oil at different rates of growth in
use, and of the possible course of oil prices
between now and 2050 given convergent policies on
the part of OPEC and the OECD[2]. If their analysis
is anywhere near the mark, oil markets are likely to
remain fragile and under strain for at least ten to
fifteen years to come. It will be virtually
impossible for OPEC to maintain oil prices without

89

the cooperation of other producers and it may even
be difficult for it to ensure price discipline among
its own members, especially if and when the Iraq/
Iran war ends.

CONCLUSION
The primary upstream structural change in the international oil industry was the elimination of the
international companies as the dominant producers of
crude oil and the consequent weakening of the
integrated structure of the industry[3]. This led to
the rapid emergence of the market for crude oil as a
major element downstream, the companies becoming net
buyers of crude. The process and its implications
have been widely discussed.

However, the structural change in the
administration of the industry itself brought about
other changes upstream which are also structural in
that they have changed, at least for the medium
term, the framework within which the industry can
expect to operate. These include a diversification
of the distribution of oil reserves as higher prices
increased the profitability of crude oil exploration
and development outside the OPEC area and also the
introduction of new technologies; they also include
a change in other conditions of supply, in line with
the objectives and politics of the OPEC countries.
The fact that the displacement of the international
companies involved the transfer of decision-making
regarding the supply and price of oil into the hands
of a coalition, and later a cartel, of governments,
and that the transfer took place under extremely
disturbed economic political conditions, created a
high degree of insecurity and uncertainty
downstream. Planning and contracts were made with
increasingly shorter horizons and spot transactions
became more and more prevalent. The ability of the
governments to maintain crude-oil prices in the face
of continuing excess productive capacity, always
precarious, seems to become increasingly so as
demand remains depressed, and in spite of the
support of OPEC efforts by non-OPEC producers.

In short, perhaps the most pervasive effect on
the downstream of the structural changes in the
upstream of the international oil industry has been
continued and omnipresent uncertainty, aggravated at
times by a feeling among consumers of dangerous
insecurity. It is this that accounts for many of
the characteristics of downstream markets and
activities.

90

NOTES AND REFERENCES

1.   This definition is not entirely applicable to the exploration stage of the oil industry in view of the uncertain nature of exploration activity.   It is always possible that in the process of normal exploratory activity oil deposits may be discovered the location and size of which are such as to constitute a genuine change in the physical framework (or structure) of the industry.

2.   See P.R. Odell and K.E. Rosing, "The Future of Oil: A Re-Evaluation", OPEC Review, Vol. VIII, No. 2 (Summer 1984), Pp. 203-228.

3.   Another upstream change, which is relevant, but which has not been considered here, is the modification of the regulatory mechanisms in the United States and the subsequent unification of the American and world markets.

7 STRUCTURE, STABILITY AND POLICY - PANEL DISCUSSION

Peter Beck, David Howell,
Walid Khadduri, Colin Robinson,
John Wiggins

The panel discussion at the end of the conference provided an opportunity for considering some wider issues linked with structural change in the oil industry. Not unnaturally the panel were asked to ponder the likely course and consequences of future structural change. Next, the panel analysed the political instability of the Gulf area as a continuing stimulus to exploration and development of alternative oil supplies. Turning to the home scene, they considered the implication of restrictions on low cost Middle East oil production for the depletion profile of United Kingdom oil reserves. Finally, attention was directed towards the related natural gas situation in the UK and whether the case against restricted supplies might also apply here. The panel consisted of Peter Beck, David Howell MP, Walid Khadduri, Colin Robinson and John Wiggins. The questions were contributed by John Mitchell (Matthew Hall Engineering), Paul Stevens, Tom Philips (Foreign and Commonwealth Office) and Ian Hargreaves (Financial Times).

What do the next ten years hold in terms of the changes in the structure of industry and in particular the relationship between the major oil companies and the nationalised oil companies in the various producing and consuming countries?

PETER BECK. I believe structural change is taking place all the time. We have not had the sort of sudden shock when the structure changes and another shock when it stops changing. The relationship between the national oil companies and the majors has changed right through time. I believe that the majors, if they are to survive, will continue to

92

co-operate with the governments and with the national oil companies. Every situation is likely to be different and if a country believes that the majors have something to offer them there will be a deal and the majors will operate in that country. If a country, rightly or wrongly, believes that it has got all the necessary experience and therefore the majors will not be needed, then there will not be a deal until those countries have different experiences, and the majors will not be able to operate. To me it is as simple as that and certainly examples from the past have indicated that. We all know of those countries which at one stage said that they did not need the majors. Some of them threw the majors out, got into difficulty and are now desperately trying to get the majors back. Other countries have operated successfully without the majors; yet others have felt that the expertise, the knowledge of the majors was needed and therefore there was co-operation again. That has been the structural pattern in the past. I do not really see that structural pattern changing. Each country will make up it's mind how it is to deal with the companies and the companies will have to make up their minds whether they wish to stay in or get out. I am not sure whether one can say very much more. We are finding, except in OPEC countries, that technology is going to get more and more difficult as it faces deep sea conditions and the need to enhance oil recovery and therefore I believe that one of the strengths of the majors is not power but knowledge. If the majors have the knowledge and are able to apply that knowledge then they can operate. If they have not got that knowledge then regrettably they will have to withdraw.

JOHN WIGGINS. There will be a need for continuing development of technology downstream as well as upstream. As the move away from heavy fuel oil continues together with the need to produce fuel with less sulphur and without any lead in it, one faces a constraint problem of how to produce products of higher quality from fewer barrels. That again will be a continuing and quite important technical change in the industry. The industry will again - and OPEC refining will sharpen this - be obliged to continue with the rather painful process of shutting down older refineries without upgrading capacity and engaging in other upgrading operations of that sort.

93

COLIN ROBINSON. The majors have proved to be
remarkably adaptable organisations and I expect they
will continue to be so. They have shown a feature
which is extraordinary to people who really believe
in economic theory because economic theory shows you
that companies always operate on minimum cost
curves. How on earth the majors have managed to cut
their costs so much in the last few years if that
were so I don't understand. Obviously costs are
very much subject to the pressure of competition and
there has been an enormous increase in the
profitability of the majors in recent times because
of these remarkable adjustments they have made. I
think that the national oil companies are here to
stay. I say that with some regret because I don't
think that most national oil companies have got
anything whatsoever to do with efficiency. They are
mostly set up for purely political reasons as far as
I can see. Finally, it seems to be very unlikely
that for ever there will be virtually no exploration
in the OPEC countries. It is an odd phenomenon
really that people have stopped exploring there when
for various political reasons they have gone into
high cost areas. Presumably one of the changes
which is going to occur someday is that people are
going to move back into these areas and exploration
is going to start. Probably at that time some more
quite low cost oil will be found.

WALID KHADDURI. I think that we will see that the
OPEC nationalised oil companies, especially the Arab
ones, will be looking carefully at the KPC position
in the Gulf and Europe and if these become
successful experiments they are likely be followed
later by expansion of upstream and hopefully of
downstream activities.

**We have heard several times today allusions to the
idea that eventually the world is going to come back
to being dependant upon oil reserves in the Gulf.
My question is in two parts. Firstly, what are the
long term prospects for socio-political stability in
the area given that one could argue that much of the
reduction in demand which we have seen since 1980
has been as a result of the OECD governments
responding, not so much to the high price, but to
political uncertainty. And then in the light of the
answer, what is the consumer government response
likely to be to this situation?**

WALID KHADDURI. I am very pessimistic about the

foreseeable future in the Middle East. I think that
what is going on in Lebanon is a sign of coming
times of communal conflict which like a cancer will
not be restricted to Lebanon. It will spread around
and will mean more trouble in many other countries
in the area. The Gulf war, the Iraq Iran war - if
it goes on as I think it will - cannot be restricted
to the Iraq/Iranian border. It would have to spill
into Gulf waters. There are several alternatives
available for both Iran and Iraq. I do not think
Kharg Island is the only alternative target that is
available and it is not the only one that would
trigger a big problem in the area. Given the
communal problem which is spreading and is really
weakening the nation states in the Arab world and is
very difficult to control together with the
insecurity that the Iranian revolution has created,
then probably OECD countries will look away from the
Gulf.

DAVID HOWELL. I wholly agree with that. Islamic
fundamentalism is going to influence the entire Gulf
area for many years to come, creating great
instability and constant doubts about the regularity
of supply from the area. This is slightly different
from saying that oil supplies would actually be cut
but the doubts will constantly be there. It is in a
sense our great defence protection because if that
fear and expectation that something terrible could
happen at any moment continue in high profile then
so also will the drive to develop alternatives and
the colossal impetus towards conservation and low
energy use. The worst thing that could happen is
for those two drives and impulses to slacken. There
is always a danger that they could rather as they
did in the late seventies. I would think that the
oil industry has to look into the Gulf area and see
nothing but instability and difficulty for years
ahead and the energy consuming nations will have to
use every ingenuity and impulse they can to develop
the alternatives and to carry on with the drive for
conservation. I think this will be on a much bigger
scale even than it has been in the last four years.

COLIN ROBINSON. I feel very pessimistic too about
the prospects in the Gulf. I keep reading stories
about how once the Iran/Iraq War is over things will
be alright, supplies will increase. I think that
this is really rather a naive view because there are
so many things that could go wrong. One must view
the future as one of considerable instability. The

problem for consuming countries though is how exactly are you going to get away from this - what do you turn to? Does Britain turn to coal for instance. That is at least as insecure. I think I would rather be dependant on oil from the Gulf than I would on coal from the British coal industry. To turn to nuclear, there is a very strong feeling against nuclear power. I am not quite sure to what extent it is a realistic option to build large amounts of nuclear power. Maybe, as David Howell says, the future is that we have just got to conserve, to try to diversify sources of supply, may be deliberately, in order to promote security. One can easily see that oil in the future is going to be a very insecure source of supply but since most other fuels are likely to be rather insecure too, it is not terribly easy to see what you do.

PETER BECK. The only hope I have is that we all really believe there is going to be a crisis and therefore there will not be one. If we really believe that by the nineties things will be alright then I am reasonably sure that there will be a crisis. As long as people really feel that supplies are very insecure all the actions are being taken to reduce oil demand in various countries. On the other hand, let us not forget that for all the instability in the Middle East so far the amount of disruption to oil has been relatively low. Perhaps there is a hope that even with continuing terrible things happening oil will continue to flow. Again when we look at the total demand for oil in the world I think we are getting too mesmerised by the OECD. It is very unlikely that the developing countries can really grow without the world having an increased oil demand - at least over the next 10-15 years, however good the new technology is which the developing world will use. Basically, I am more of an optimist largely because all the signs are that we ought to be pessimistic.

JOHN WIGGINS. I do not want to understate the importance of the Iran/Iraq war - it obviously is a very terrible thing and it's very much to be hoped that it will come to an end as soon as possible. But it seems to me to be taking an unduly alarmist view to suppose that that particular conflict is likely to spread all down the south side of the Gulf. As I understand the structure of the societies on the south side of the Gulf they are very different from either Iran or Iraq and maybe have some defence from their own versions of Islamic fundamentalism which they have always operated. The other thing I would want to say is that even Iran, revolutionary Iran, needs money and exports oil. Iraq has had it's output cut by 2/3rd as a result of the war. When the war stops they will need money urgently. I think that the impetus to conservation has depended very much on the high price and even after two years' of slightly beguiling real prices oil is still expensive compared with what it was in the fifties and sixties. Ordinary economic forces will help to protect us to a significant extent. The overhang of Russian gas may help the industrial countries a bit. Lastly. the oil producers with really very large revenues in relation to their populations have a very strong mutual interest with the industrial countries in monitoring the stability of the present free world economic system. The State of Kuwait, as I understand it, already gets as much GNP for income in terms of income from abroad on investments it has already made as it does from current production of oil, and I would expect Saudi Arabia to move in that direction through the remainder of this century. There are therefore quite a lot of forces that still tell in favour of sensible world economic co-operation continuing.

WALID KHADDURI. There are really two conflicts here. There is the communal one (the problem of a non homogeneous population) and the question of intervention in another person's country, particularly in Iraq/Iran. The communal conflict is just as dangerous as the other because if this spreads and there are signs that it is spreading into other countries, it will threaten the nation states in the area. And it will be a source of instability which will affect the decision making process, the interests and the economic allocations and decisions and investments. The other point is that the Iraq/Iran War continues. I think Iraq eventually will not allow this kind of war to take place with joint fighting on its borders. It will try to attack more serious, more strategic Iranian targets to try to put an end to it, and that will escalate things. It doesn't mean that targets have to be on the eastern or the western shores of the Arabian Guilf, but it could affect strategic industries, oil industries which would escalate the conflict. Oil flow may or may not be disrupted. The security of the area and the way that other governments, international companies and the people

of the area themselves look on the region as such
will definitely be changed. The whole perception of
the area as a source of investment and wealth for
it's people and for those working with them will
definitely be changed, if this insecurity
continues. This is really a longer term concept
than the immediate security of oil flow. I am not
so much worried about Islamic fundamentalism as such
- it is one force in the area. The communal
conflict is much worse than that and is much more
difficult to contain because Islamic fundamentalism
exists, for example, in Egypt but Egypt does not
fact the communal problem. Further problems are
caused by the Arab Israeli War and the inability of
the U.S. to do anything really while trying to
maintain superiority.

**We have heard a lot today about long term prospects
for Middle East reserves. Do the panel think that
United Kingdom production policy is about right? If
U.K. production is in the end going to decline then
should we adopt a different outlook on depletion
policy?**

JOHN WIGGINS. Ought we to postpone our production?
The first thing to be said about that is that the
U.K. is not actually an isolated, insulated economy
without any interest in what happens to the
economies of other industrial countries. In
particular we export 60 percent of the oil that we
currently produce and our refineries are only
dependant as to about 60% on U.K. Continental
Shelf. We do still use and process a good deal of
heavier Saudi and other crudes which are cheaper
when it is to the advantage of our own national
income to do that. So I think it is, in a sense, a
misleading idea to suppose that we could somehow or
other keep our oil in the ground and then at a later
point - assuming we are still in the European
Community and the I.E.A. and such like organisations
- produce it when life got difficult and keep it all
for ourselves. If you advocate a policy which says
you ought to keep your oil in the ground other than
for some sort of strategic reason, you are thereby
saying that the oil will increase in price in real
terms by more than enough to compensate us for not
extracting the resources in the meanwhile which we
could invest in other productive use. I personally
don't feel that I have any assurance that oil really
will be more expensive in real terms. It would need
to be a good deal more expensive - $50+ pb at least

by the year 2000. Looking at things that have been happening in the world energy scene recently I am personally impressed by the overhang of gas supplies as well as oil supplies and that seems to me to be a factor which would lead oil prices to be a bit lower in the 1990s than might otherwise be expected. So from that point of view I think our policy is about right. So far as our fiscal policy is concerned obviously the Department of Energy has played some role in recent discussions within government about what ought to be done, so that I bear some very small responsibility for that. Again I think the policy is sensible. What we have done is to organise the tax system so that where you can produce oil profitably taxes don't discourage you from doing that.

DAVID HOWELL. The tax system is itself an expression of public policies about the health and mortality of North Sea oil production. A minister has pointed out recently that the tax changes in one of the recent budgets had had a positive effect in accelerating exploration and development. So there is an example of a government policy change actually influencing the pace of activity in the North Sea, the pace of production as well as exploration and development. But that was done - not for any reason associated with a view about the depletion profile, but really as part of a broader economic goal of wanting more economic activity in an industry that felt it had the tax system set against it through past changes and was saying that it was going to create problems in the future. I think that the present policy is the only appropriate one because the economic case for depletion is an almost impossible one for governments to judge and assess. Who is possibly to know with this vast spread of guesses about what the price is going to be? Assessments by government let alone by businessmen about the long term values of keeping oil in the ground economically are almost impossible to make. In as far as one can make such assessments they seem to me to point very much in the direction of producing and collecting the cash now and maybe reinvesting it in one way or another, rather than making some wild investment decision, highly speculative about future prices, involving keeping the oil in the ground. That's quite aside from government powers to carry out such policies - there are certain powers but they are not probably fully in place to operate a full blown depletion policy as

of now. As for the strategic argument, I think it
is a fallacy to believe that retaining supplies of
oil on one's continental shelf in our kind of
economy is necessarily the best form of security of
supply. On the contrary - stealing a line from
Colin Robinson here because he has put these things
much more clearly than I have - the more diverse
one's sources of supply not only of oil but other
energies probably the more secure one is. The
arguments for keeping oil in the ground for security
of supply reasons in an industrial country which is
also a big oil producer such as ourselves - and we
are nearly unique in this role amongst the medium
sized economies - I think are very weak indeed. As
for the rest of our policy just let me say a word
about our policy on production generally in relation
to world oil price and OPEC. There was a good deal
of comment in March 1983 about the U.K. government's
attitude to world crude oil prices and about the
urging from OPEC that a position should be taken by
HMG or by its agent the old trading arm of BNOC, as
it existed then, in relation to prices. I suppose
there was a government view and I suppose it remains
that, on the whole, violent movements in price,
whether up or down are bad news. Certainly that was
the view taken in March 1983, not merely for oil
production reasons or because the Treasury wanted
the oil revenue but because it would lead to all
kinds of repercussions world wide in the banking
system and in the international financial
community. So a view was taken that while no one
ruled out the proposition that in the longer term a
nation like ours would like to see a lower worldwide
oil price, nevertheless a very precipitous move
would also be bad news and that therefore there was
something to be said for being as sympathetic as our
kind of oil regime can be to any necessary moves and
pressures to prevent the oil price breaking. Was
that right or not? I think probably in the very
short term it was right. Our long term interest in
a much lower oil price is much stronger than perhaps
people realise, despite the fact that we are the
world's fifth largest oil producer at the moment.
And as for the consumption side, we have always
worked with IEA, taking the lead role as a consumer
country and I think our position there as primarily
a consumer acting with the other consumer countries
and ready to operate in the emergency procedures and
to draw down stocks and so on is right.

PETER BECK. I'd like to come to the other side of

the question, namely will a changed depletion policy improve the situation in the U.K. in the nineties. I really doubt it. Such a debate occurred in 1979/80 and the general conclusion was for "no particular depletion". Really it is a bit late to have that particular debate again because our production is about to peak in a couple of years' time. If we are now going to shade it off irrespective of the short term factors the amount of additional oil we will have in the 1990's is really miniscule - if spread over the nineties. We know that production from the North Sea will be substantially less in the nineties than it is now, as one of the few facts about the future, largely because the oil companies have not got the means of developing sufficient reserves in the period available to overcome the problem of lower production from the very large fields. So we shall be either importing small quantities or be self sufficient. But I believe that the world as a whole will again be far more dependant on OPEC oil and largely in fact on Gulf oil than it is now and whether we deplete our oil more or less now will not resolve that problem one way or the other.

COLIN ROBINSON. I agree, on the whole, with what has been said in support of present policy and I would perhaps put it just slightly differently. First of all I agree with what David Howell said about the tremendous uncertainties. If you are going to use a policy of deferring production initially you have to have quite a good idea of what would happen if you did not do anything. That is quite difficult at the moment because there is much uncertainty about the future of North Sea production - the range is tremendous. There is actually quite a good theoretical case for a certain amount of production deferral even though with oil prices rising relatively slowly and any likely discount rate such a policy would lead to a lower present value for the North Sea fields. You could argue that there would be some security gain so that in effect the market price is undervaluing the real value of the oil. You might get some protection against the possibility of very big oil price increases in the future, a kind of insurance policy. Again, it is possible that you could smooth out the re-entry problem when the U.K. moves from being a net exporter into being a net importer. Another argument is that oil companies use higher discount rates than the social discount rate. The

101

problem however is that if you embark on this kind
of policy what you are actually going to do is to
take it out of the hands of an imperfect market and
put it into the hands of an imperfect government.
Once you have depletion control legislation there is
absolutely no telling what is going to happen. I
personally believe that most governments in Britain
are likely to have a built-in interest in
accelerating depletion if possible because they want
to get revenue in quickly, and although I can see
that there is a theoretical case for a sort of
flattening out of the lump. I think if you once
embarked on that policy, it would probably get
twisted so that it became something quite different.

IAN SEYMOUR. I think that there is another aspect
to the production question with regard to the North
Sea. There are times of extreme market weakness as
in 1983, where it would be extremely useful both for
the U.K. and other exporters, possibly the world as
well, if the U.K. government had the power at least
to cutback production temporarily without at all
altering it's long term depletion policy. There are
crucial times in the market where the ability just
to hold back production is very, very useful, and it
is regrettable that the U.K. government does not
quite have the instruments in place for it to do
this.

DAVID HOWELL. If there is a time of market weakness
why isn't it in the interest of the producers and
the companies to regulate production in as far as
they can? Obviously in the very short term the
technical and engineering constraints on what can be
done in the way of short term production regulation
are really rather small - certainly what can be done
without very serious economic damage overriding any
economic damage you may be doing by producing at a
higher level but selling into a weak market.
However, the government's view all along has been
firstly that a great deal of the investment in the
North Sea's entire regime was built upon certain
very clearly given undertakings like the Varley
undertakings, and no more than that, and therefore
investment was undertaken with very precise
calculations based on those undertakings and they
have been adhered to. Secondly, the view has been
taken that if there are going to be economic forces
at work discouraging production at times of weak
markets those forces will work anyway. It doesn't
need a government official to rush in and tell the

company that it's not good news to be in producing at such a rip roaring rate when the price is falling like a stone. That's the reason why an elaborate system of powers for actual production control (as opposed to development delays) and trying to smooth out the lumps by depletion policy have not been put in place.

COLIN ROBINSON. Actually, if you look at the legislation it's perfectly possible for the British government to do almost what it wants with production. It has denied itself the powers through the Varley assurances which have been extended, but if you look at the legislation itself - the Submarine Pipelines Act - it is in fact quite possible. Presumably from the end of 1984 if the government did not want to review those assurances it could regulate production - I'm not suggesting it should because I don't think it should, but it would be possible to do so under the legislation.

PETER BECK. You can leave the oil companies out of this issue because first of all we have to agree production profiles with the government. Secondly, for companies really to cut back must be the very last resort because of course the marginal cost of producing oil is extremely low indeed once you've sunk the investment. If you reduce production you really reduce the cash flow for this year. So for the companies to decide for some longer term motives to cut back on their own - even if they could get government permission to do that - is most unlikely.

JOHN WIGGINS. Looking back at the history of the last 12 or 15 years we had a period when OPEC could say quite reasonably that the industrial countries' growth depleted their resources too quickly. What they did was to put prices up, and what the industrial countries including the U.K. did was to invest in their own production rather than depend on OPEC. The world outside the Communist area can be divided into two groups - one group is a group of countries who broadly speaking belong to OPEC and operate with control of production in the hands of state companies and this group now has to take collective responsibility for setting the broad price of crude oil. The rest of the world including the United State and Canada are investing in producing within the economic framework which has been created by the OPEC decisions of various kinds. I do not believe that in the U.K. we could start to restrict production as part of a deal with OPEC unless this was done as part of a world monetary agreement, and this I do not believe is possible. If we simply did it ourselves without any solid broader international agreement I would expect that the companies would become extremely frightened of British government. The prospects for continuing success in exploration and development of the North Sea to which the government necessarily attaches a lot of importance would deteriorate very sharply.

Does the argument that there is no case for hoarding oil and restricting trade in it, apply in the case of gas as well as oil?

DAVID HOWELL. We are getting within sight of the point which would be a great relief for politicians when gas prices could be settled by market forces by which I mean a point at which the North Sea producer would be free to pipe gas to the housewives in Holland, West Germany and France as well as to the housewives of the U.K. I know that BGC doesn't share that view at all and so long as our industrial gas prices were about the same as in the rest of Northern Europe whilst our domestic gas prices were only 1/3 to one half the corresponding levels an enormous problem existed. Things have changed. The ceiling on gas prices has been removed, some decontrol element has taken place over the years. We are told, and correctly I think, that it's still not enough. Indeed, the price as you know is still something like 44 pence per therm to our houses and about the same into Holland but about 66 pence per therm into Northern Germany and about 58 pence per therm into France and so there is still a gap. But it is closing all the time and once it does close I see no reason why one should not liberate North Sea gas production. Then we would have a market price and politicians would cease to have to wear tin helmets every time they talked about domestic gas prices. That will be a great advance, and would be very healthy for North Sea gas development as well.

COLIN ROBINSON. I agree very much with what David said. We have a very odd policy for gas in this country. We don't allow exports but we do allow imports, but the imports have to be sold by an organisation which has a monopoly of the markets - a very odd kind of policy. It's a price depressing policy in the domestic market and it clearly had a very bad effect on exploration and development in the Southern Basin of the North Sea. The time has certainly come now where it is important to establish a proper market in gas which you can only do by allowing exports to take place. The direct industrial sales legislation in the Oil and Gas Enterprise Act was helpful in that it probably induced BGC to bid a bit more for gas and so create the potential competition which raised the price. But the only way to get to a proper market in gas is to allow exports, and there seems to be no reason why one shouldn't do that. Once you get a pipeline connection to the continent you are then going to have a wider range of gas imports too rather than just being dependant on Norway, Holland or the Russians for that matter.

PETER BECK. We do need some mechanism through which the continental gas pricing which indeed is a market pricing system can be translated to this country. If not the danger is that for one reason or another we start developing gas at a cost which is far higher than the country need pay to get its supplies from say the Dutch or the Norwegians. Alternatively, for the reasons mentioned, if there is a monopsony in this country, then it might turn on the tap, say we need more gas then suddenly close it and say 'no we don't' and that creates terrific swings in exploration and production developments which really can be very serious.

JOHN WIGGINS. Gas is a rather odd commodity that cannot be transported around in anything like as convenient or flexible a way as oil products. Once you are tied into a commitment to a particular gas stream you are tied into it permanently and there is much less flexibility of moving it about. I think that an uncontrolled export regime would still leave the government with a responsibility to impose quite strong controls so as to ensure that the U.K. actually got the benefit from the sale of the gas and that it was not appropriated by the licensee selling it to foreign purchasors at relatively cheap prices.

8 SURVEY OF OIL PRICE EXPECTATIONS
22/23 March 1984

David Hawdon

This survey of oil price expectations was carried
out during the international conference on the
Changing Structure of the World Oil Industry held at
the University of Surrey on 22nd and 23rd March
1984. The respondents were economists, managers and
planners working either directly for commercial
energy companies, energy consultancies, or
government departments involved with energy policy
together with academic economists actively involved
in the analysis of oil industry problems. In total
56 responses were obtained out of 78 questionnaires
circulated of which 30 were from industry 6 from
government and 20 from academics.
    Before discussing the results it is important
to bear in mind certain potential sources of bias.
In the first place the sample is obviously not
random and results can not therefore be subject to
rigorous statistical analysis. This lack of
randomness is reflected in the high proportion of
academics and the low proportion of government
economists represented. We have tried to allow for
this by presenting separate results for industry,
government and academics and invite the reader to
apply his own weighting if it is felt necessary.
Secondly, the survey was carried out over a very
short space of time and none of the respondents was
able to carry out any detailed analysis of the
issues in advance. Thus the results reflect very
much the 'snap' judgement of a group of senior
economists. They give a 'feel' for the anticipated
crude oil price rather than a prediction based on
any sophisticated forecasting exercise.

The Questionnaire (see Appendix)
Respondents were asked three questions. The first
was intended to elicit a view concerning the likely

price of crude oil (OPEC marker) in 12 months'
time. We specified that the answers should be in
terms of "money of the day". This means of course
that people were asked implicity to make all kinds
of judgments about dollar inflation rates as well as
about movements in the real price of oil. Phrasing
the question in this way does, on the other hand,
yield estimates which are directly comparable with
the eventual outcome and we hope to make this
comparison in due course. It may be argued that
this type of question requires more than could
reasonably be expected of participants at a
conference, without access to their full company
backup resources. We tried therefore to avoid too
great an air of precision by asking for price ranges
(less than $25, $25-27, $28-30, $31-34 and over $34)
rather than specific prices. Nevertheless it is
possible that respondents might err on the side of
caution in completing both this question and the
second one which asks them to express a view about
the likely price 5 years' hence.
      The third question was designed to discover
the major reasons behind the price judgment made.
We asked respondents to nominate the four most
important factors which they anticipated might
affect the price of crude oil over the next 5
years. Six specific factors were listed but
respondents were given the opportunity to identify
other factors which they believed were important.
In analysing the answers we constructed an
"importance rating" statistic for each factor by
summing the values (1 to 4 for the four most
important factors; and 5 for the remaining three
factors) and then dividing by the total number of
respondents. This procedure yields a
straightforward ordering of factors although, as
usual in economics, cardinal comparisons should not
be made since we asked respondents only to rank the
factors and not to indicate their strength of
feeling about the importance of each one.

The Results
1.   The Likely Price of Crude Oil in 12 Months'
        Time (see Table 1)
Two clear results emerged from the analysis of
responses - in the first place over 3/4 (44 out of
56) of the respondents felt that price would be
within the $28-30 range in 12 months' time and
secondly, no one believed that prices would either
fall below $25 per barrel or rise above $34 in the
period. There was no significant difference between

the industrial and academic respondents on this
issue. The government economists agreed unanimously
on the $28-30 price range although the numbers were
too small for any real statistical comparisons. Not
surprisingly each category had a median price
expectation of $28-30 pb.

2.   The Likely Price of Crude Oil in 5 Years' Time
      (see Table 2)
Perhaps not unexpectedly, a much greater diversity
of views about the 5 year ahead price was held by
our sample. There was, however, a unique modal
choice of $31-34 pb by over half of the respondents
(31 out of 56 or 55.4%) and this range was also the
median range. It represents a price on average 12%
higher than in 1984 implying an expected compound
growth of just over 2% per annum to 1989, in money
of the day terms.
        There was some disagreement between the
academic and industrial economists on how the price
would change. Whilst the majority in both
categories opted for $31-34 pb, a greater proportion
of the academics (30%) regarded a lower price as
likely (both compared with those academics who voted
for a higher price (30% compared with 10%) and also
in comparison with the industrialists, 16.7% of whom
felt that prices would be lower than $31-34 pb and
26.7% higher. None of the government respondents,
interestingly enough, believed that prices would
fall below current levels and the response was split
evenly between those anticipating no change in
prices over the period and those expecting prices to
rise.

3.   The Important Factors   (see Table 3)
Two factors in particular are seen as having an
overriding significance for the determination of oil
prices over the next five years - OPEC's ability to
control oil production on the supply side - with a
score of 1.9 - and economic growth in the OECD
countries on the demand side with a score of 2.3.
Slightly greater weight is placed on OPEC's power of
control than might seem warranted by recent history
although this may be influenced by OPEC's success in
maintaining the marker at $29 pb since January
1983. The third most important factor is seen to be
the Iran/Iraq war, but here unfortunately the
phrasing of the question does not permit inferences
about the direction of this effect. The remaining
factors - energy conservation, changes in other
energy prices and developing countries' financial

problems - are not believed likely to affect prices
very much in the five year period.

Why then did the majority of the sample
believe that oil prices will rise to $31-34 pb by
1989? This question can be answered, partially at
least, by comparing the "importance ratings" of the
group which anticipated increased price with the
'importance ratings' of those who did not. The
respondents were divided into two subsamples - those
voting for $31-$34 or higher and those opting for
$28-30 or lower prices over the five year period -
and the importance ratings were recalculated for
each subsample. The results of this exercise (given
in Table 3) show, interestingly enough, that both
groups rank the factors in the same order of
importance. The small differences between ratings
suggest however that the higher price prediction is
associated with greater weight being given to
economic growth in the OECD, to the effects of the
Iran/Iraq war and to other factors, especially
political uncertainty and "the unexpected"
(specifically mentioned under "other" - see
Table 4). The greater weight given by this group to
Energy Conservation is more difficult to understand,
unless higher prices are seen as necessary to
support existing conservation investments. Finally,
factors such as the financial problems of the
developing countries, the production control of OPEC
and changes in other energy prices are seen as less
important by this group.

ACKNOWLEDGEMENTS
The author gratefully acknowledges the assistance of
Jacqueline Read in processing the survey results on
the University of Surrey's PRIME computer system.
Paul Stevens and Colin Robinson gave much helpful
advice during the preparation of the questionnaire.

Table 1

What is the world price of crude oil (OPEC marker presently $29 p/b) likely to be in 12 months' time in money of the day?

$ p/b	Under 25	25-27	28-30	31-34	Over 34	TOTAL
ALL	0 (0)	6 (10.7)	44 (78.6)	6 (10.7)	0 (0)	56 (100)
Academic	0 (0)	3 (15.0)	15 (75.0)	2 (10)	0 (0)	20 (100)
Gov't	0 (0)	0 (0)	6 (100)	0 (0)	0 (0)	6 (100)
Industry	0 (0)	3 (10.0)	23 (76.7)	4 (13.3)	0 (0)	30 (100)

(%)   Median value $28 - 30

Table 2

What is the world price of crude oil (OPEC marker) likely to be in 5 years' time in money of the day?

$ p/b	Under 25	25-27	28-30	31-34	Over 34	TOTAL
ALL	1 (1.8)	3 (5.4)	10 (17.9)	31 (55.4)	11 (19.6)	56 (100)
Academic	1 (5.0)	2 (10.0)	3 (15.0)	12 (60.0)	2 (10.0)	20 (100)
Gov't	0 (0)	0 (0)	3 (50)	2 (33.3)	1 (16.7)	6 (100)
Industry	0 (0)	1 (3.3)	4 (13.3)	17 (56.7)	8 (26.7)	30 (100)

(%)   Median value $31 - 34

Table 3

What are the four most important factors which will influence
the world price of oil over the next 5 years?

Importance Ratings	Total Sample (56)	Respondents choosing	
		< $30 (14)	> $30 (42)
OPEC's ability to control production	1.9	1.6	2.0
Economic Growth in OECD	2.3	2.4	2.3
The Iran/Iraq war	3.6	3.7	3.5
Energy conservation	4.0	4.2	3.9
Financial problems of DC	4.3	4.1	4.3
Changes in other energy prices	4.3	4.2	4.3

(sample size)

Table 4

Other Factors Influencing the Price of Oil

	Number of Mentions
The unexpected	3
LDC demand growth	2
Gas supplies and prices	1
Perceived scarcity	1
Non OPEC production	1

APPENDIX

SEEC CONFERENCE SURVEY 1984

In view of the high level of expertise of those attending the SEEC conference, the Centre felt it would be interesting to run the following brief questionnaire to see if any broad consensus existed on the future of the oil market. It would be greatly appreciated if those attending could answer the questions below and **return the form before 11.30 a.m. on Friday** to one of the boxes which will be located in the conference hall. Anonymity is of course guaranteed **(DO NOT SIGN THE PAPER)** and the aggregated results will be presented to the conference before the closing panel discussion.

1   What is the world price of crude oil (OPEC marker presently $29 p/b) likely to be in 12 months' time in money of the day? (Please tick).

$ p/b	Under 25	25–27	28–30	31–34	Over 34

2   What is the world price of crude oil (OPEC marker) likely to be in 5 years' time in money of the day? (Please tick).

$ p/b	Under 25	25–28	28–30	30–34	Over 34

3   Which are the four most important factors which will influence the world price of oil over the next five years? (Please order 1 – 4 with 1 as the most important, 2 as the next most important, etc.).

Economic growth in the OECD

Financial problems in the developing countries

The Iran/Iraq war

OPEC's ability to control production

Changes in other energy prices

Energy conservation

Other – Please specify:-

4   Please indicate whether you are – Academic

– Government

– Industry/Commerce

Printed and bound by CPI Group (UK) Ltd, Croydon, CR0 4YY

21/10/2024

01777088-0019